BIG BANG

From Myths to Model

Revised First Edition

By Jason P. Smolinski
Calvin College

cognella® ACADEMIC PUBLISHING

Bassim Hamadeh, CEO and Publisher
Michael Simpson, Vice President of Acquisitions
Jamie Giganti, Managing Editor
Jess Estrella, Senior Graphic Designer
Carrie Montoya, Manager, Revisions and Author Support
Natalie Lakosil, Licensing Manager
Kaela Martin and Berenice Quirino, Associate Editors
Mandy Licata, Interior Designer

Copyright © 2017 by Cognella, Inc. All rights reserved. No part of this publication may be reprinted, reproduced, transmitted, or utilized in any form or by any electronic, mechanical, or other means, now known or hereafter invented, including photocopying, microfilming, and recording, or in any information retrieval system without the written permission of Cognella, Inc.

Trademark Notice: Product or corporate names may be trademarks or registered trademarks, and are used only for identification and explanation without intent to infringe.

Cover image copyright © User:Superborsuk (CC BY-SA 3.0) at http://commons.wikimedia.org/wiki/File:Gravitation_space_source.png; © Depositphotos/prill; © Depositphotos/cherezoff; © /agsandrew.
Printed in the United States of America

ISBN:978-1-5165-1181-5 (pbk)/ 978-1-5165-1182-2 (br)

Contents

ACKNOWLEDGEMENTS — IX

PREFACE AND INTRODUCTION — XI

1. ASTRONOMY: ONE OF THE FIRST SCIENCES
1.1. What motivates us to want to learn about astronomy? — 1
1.2. How big are the stars? — 3
1.3. How far away are the galaxies? — 4
1.4. What does the Universe look like? — 8
1.5. What is the age of the Universe? — 10
1.6. How confident are we in our knowledge? — 12

2. ANCIENT IDEAS
2.1. Cosmology of ancient civilizations — 19
2.2. Ancient sky watchers — 24
2.3. From supernaturalism to natural philosophy — 25
2.4. Early thinkers — 26
2.5. The geocentric universe — 32

3. RENAISSANCE ASTRONOMY
3.1. The Copernican "revolution" — 39
3.2. Refinement of the model — 44
3.3. Galileo and his telescope — 51

4. THE NEWTONIAN REVOLUTION
4.1. Isaac Newton — 57
4.2. Universal gravitation — 58
4.3. Additional contributions — 62

5. THE DAWN OF MODERN PHYSICS
5.1. A shift in perspective — 67
5.2. The special theory of relativity — 69
5.3. Generalizing relativity — 77
5.4. Defining a new normal — 82

6. LIGHT AND MATTER
6.1. The nature of light — 85
6.2. The matter of little things — 89
6.3. Interaction between light and matter — 95

7. MEASURING THE STARS
7.1. The role of light in astronomy — 107
7.2. Telescopes — 108
7.3. Astronomical imaging devices — 116
7.4. Brightness measurement — 117
7.5. Spectroscopy — 120

8. THE EXPANDING UNIVERSE

8.1. "The Great Debate"   125
8.2. Edwin Hubble   130
8.3. The "stretching" universe   132
8.4. A new paradigm   140

9. THE BIG BANG MODEL

9.1. A rival model arises   143
9.2. The big bang model develops   145
9.3. Observational evidence—the cosmic microwave background   148
9.4. The origin of matter   155
9.5. The model as we know it   163
9.6. Problems and solutions   167

10. MODERN COSMOLOGY

10.1. Accelerating expansion   173
10.2. Cosmic accounting   177
10.3. What may have caused the Big Bang?   180
10.4. Concluding remarks   182

REFERENCES   185

INDEX   193

Dedication

To the student of life: never stop seeking, pursuing, and learning.

Acknowledgements

I would like to thank my wife, Ashley, for humoring me as I set up a very cluttered work table in our living room. Your love and support throughout this process has meant more to me than you know. Additionally, the enthusiasm of my family and friends has been a consistent source of motivation. Finally, my departmental colleagues also deserve a great deal of thanks for supporting me when others may have felt that there were better ways to spend my time and energy. Thank you all; I am in your debt.

Preface and Introduction

The historical progression of astronomy is just as fascinating as the scientific progression of astronomy. The genesis of this textbook rose out of the desire for a book that not only looked at what we currently understand about our Universe and why, but also how we arrived here. Investigating the human side of science—when and how ideas arose, and what sort of resistance existed to these new ideas—provides the student with a fuller appreciation for science as it really is: messy.

The discussion of origins can sometimes instigate arguments over philosophical and/or religious perspectives. This textbook is intended to be religiously neutral and neither attacks nor advocates any religious views, but instead seeks to teach the scientific consensus as it currently stands. To this end, the book's title *Big Bang: From Myths to Model* uses the word 'myths' not to connote false ideas but instead to indicate that the progression of our understanding of the Universe finds its beginnings in the lore of the ancients. The word derives from the Greek word *mythos*, which refers to a story or set of stories that were passed on verbally from one generation to the next and that had a significant meaning for a particular people group. The word's origin does not itself indicate truth versus fallacy, and those aspects of any ancient story are not debated in this book either.

This textbook is designed and intended to serve an introductory-level college course oriented around a historical and scientific treatment of the development of insights into the Universe around us. It is organized in as close to chronological order as possible while attempting to tie relevant details together when they are needed. No prior knowledge of astronomy is assumed, and the mathematical applications are kept to a minimum so as not to overshadow the scientific concepts.

Typical astronomy topics like lunar phases and star formation/evolution are omitted because they do not directly and substantially pertain to our understanding of how the Universe began, although it is possible that this claim could be debated. Other topics, such as Doppler shift and parallax, are introduced on an "as-needed, when-needed" basis. Rather than devoting one or more chapters to prepping the reader for what is to come much later, an attempt has been made instead to address the needed details and concepts at the moment they are helpful to the reader. The author's goal was not to produce a comprehensive introductory astronomy textbook, but one that focused on the story surrounding how we have come to believe what we do about the Universe's origins, and the scientific details therein.

In reading this book, it is the author's hope that the reader will recognize science for both its strengths and limitations. Science is more than just a collection of facts, and indeed the scientific consensus regarding these facts can change. An effort has been made to avoid speaking dogmatically, instead emphasizing that much of what is contained in this book is, or explains, what is currently believed. An intellectually honest assessment of the fluidity of science should justify this approach.

CHAPTER ONE

A Print Of The Copperplate Engraving For Johann Bayer's *Uranometria* (1661) Showing The Constellation Orion.

United States Naval Observatory Library, "Uranometria Orion," http://commons.wikimedia.org/wiki/File:Uranometria_orion.jpg. Copyright in the Public Domain.

Astronomy

One of the First Sciences

Learning Objectives

In this chapter you will learn:
- » How scientists develop ways of explaining the world around us
- » What fundamental structures comprise our Universe
- » How to express very large or very small numbers in a convenient notation
- » Sizes and distances within our Universe

Key Words

- Light-year
- Galaxy
- Metals
- Globular Cluster
- Science
- Hypothesis
- Theory

1.1 What motivates us to want to learn about astronomy?

For thousands of years, humanity has gazed up at the night sky and reveled in its beauty. With its countless points of light and mottled fuzzy features, it is impossible not to gaze in awe and ponder what one is seeing. Where did it all come from? What are those tiny twinkling specks that litter the sky? Why are some brighter than others? What produces that great nebulous pathway overhead? **Figure 1.1** provides an example of the view seen by countless generations.

2 | Big Bang: From Myths to Model

Figure 1.1. A meteor streaks across the Milky Way arching over the Very Large Telescope, in northern Chile.
Copyright © ESO/S. Guisard (CC BY-SA 3.0) at http://commons.wikimedia.org/wiki/File:The_2010_Perseids_over_the_VLT.jpg.

As our society has grown and developed, this type of philosophical rumination has evolved into today's modern scientific inquiry, where humans have not only assembled a vast array of questions but have also developed the means to investigate such questions. This has provided us with a wealth of information that has been used to give insight into, and in many cases answers for, some of these inquiries. However, it is not always clear to many of us how scientists went from question to data to conclusion. For example, how did astronomers start from the Earth-based view of the Milky Way shown in **Figure 1.1** and develop the current model of our Milky Way Galaxy shown in **Figure 1.2**?

Figure 1.2. An artist's impression of the overhead view of our Milky Way, a spiral galaxy.
NASA JPL, "Milky Way Galaxy," http://www.nasa.gov/images/content/188404main_hurt_Milky_Way_2005-590_lg.jpg. Copyright in the Public Domain.

In this chapter, and the chapters that follow, we will look at some of the most pressing and persistent questions that humanity has faced—not only those posed above, but some that delve deeper into our psyche: where did the Universe come from? Why are things the way they are? What does the future hold? Since many of these questions have fairly high-level answers, we will build a foundation of knowledge along the way that will allow us to consider some of these topics from a scientific perspective. As we begin, you can increase your engagement with the reading by writing down your own fundamental questions. Make a list and consult it regularly as you read, jotting down the answers and explanations as you arrive at them. What questions do you hope to have answered?

1.2 How big are the stars?

Part of the intrigue about astronomy stems from the fact that we really have nothing to relate it to here on Earth. As a field, astronomy deals with objects that are the largest, farthest, oldest, and fastest. It is this aspect that stimulates the imagination.

- On Earth, if something is "really old" it might be thousands or millions of years old. On astronomical timescales this is a mere drop in the bucket, with "really old" objects being greater than 10 *billion* years old.
- On Earth, if something is "really big" it might weigh several tons or be hundreds or thousands of feet tall. When astronomers speak of something "really big" it might be anywhere from hundreds of times the mass of the Sun to billions of solar masses, and thousands to millions of *light-years* in size.
- On Earth, if something is moving "really fast" it might be traveling anywhere from 100 miles per hour (45 meters per second) to hundreds of meters per second (e.g., a speeding bullet). However, in space our planet orbits the Sun at a speed of 30 *kilometers* per second (that's 30,000 meters per second, or roughly 67,100 miles per hour!). The Sun orbits the center of our Galaxy almost ten times faster. Also, it is not unusual to find galaxies that are moving within galaxy clusters at speeds of 500–1,000 kilometers per second or faster! By comparison, that's over 30,000 times faster than the typical highway speed of a moving vehicle.

With all of these staggering values, how can our minds even begin to fathom the dimensions of the Universe? We do so by creating analogies. For example, suppose we were trying to conceive of the size of our Sun, shown in **Figure 1.3**. If we could shrink the Sun down to be the current size of Earth, then the analogous Earth would be approximately 73 miles in diameter—roughly the distance between New York City and Philadelphia. An average person on this resized Earth

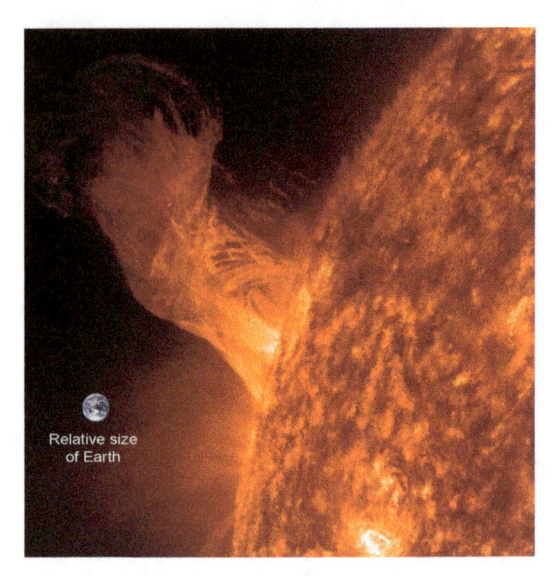

Figure 1.3. The Sun, our own personal star, with the Earth to scale.

NASA, SDO, and Steele Hill, "The Sun," http://www.nasa.gov/mission_pages/sdo/news/solar-ballet2.html. Copyright in the Public Domain.

would be about the size of your thumbnail. Comparing your thumbnail to the size of the Earth gives you an idea about how you compare to the Sun.

However, the Sun is a relatively average star. If our Sun were scaled down to be the size of our Earth, then the largest stars would (on our revised size scale) still end up being the size of our current Sun—and larger! This means you can look at **Figure 1.3** and replace the image of the Earth with an image of the Sun, and the enormous star next to it would represent the low end of the largest stars in our Galaxy.

On the other hand, there are also stars much smaller than our Sun. Ultimately, the maximum size a star can have is related to its mass, its rotational speed, and its composition. On the low end, the minimum size a star can have is that which still contains enough mass to provide the internal pressures and temperatures needed to maintain nuclear fusion in the core. This lower limit in mass is at approximately 8% of the Sun's mass (or, as astronomers write it, 0.08 $M_\odot$, where the small subscript symbol represents the Sun), which is about 80 times the mass of Jupiter. Interestingly, an object with this mass is also comparable in size to Jupiter, being only slightly larger in radius than our largest planet.

In our revised size scale from earlier, if we shrank our Sun down to be the current size of Earth, the smallest stars would span slightly less than one-half the width of the United States. Comparing the actual sizes of the states of Alaska or Texas to the Sun gives you an idea about how the smallest stars compare to the largest stars.

1.3 How far away are the galaxies?

We all have a sense that the Universe is a big place. The Earth, with its diameter of roughly 8,000 miles, seems big enough by our experience. The distance from Earth to the Sun, however, is over ten thousand times that number—enough to line up 100 Suns side by side. This distance is referred to as an **astronomical unit (AU)** and

Planet Name	Distance from Sun	
	(10^6 mi)	(AU)
Mercury	36.0	0.39
Venus	67.2	0.72
Earth	93.0	1.00
Mars	141.6	1.52
Jupiter	483.6	5.20
Saturn	886.5	9.54
Uranus	1783.7	19.19
Neptune	2795.2	30.07

Table 1.1. Planetary distances from the Sun, in millions of miles and astronomical units.

in 2012 was defined by the International Astronomical Union to be equal to 149,597,870.7 kilometers, or approximately 93 million miles. By this definition, the Sun is 1 AU away.

On the scale of our solar system, Earth is actually quite close to the Sun. Suppose we shrank the solar system down so that it could fit across the width of this page. If the Sun was on the left side, and Neptune was all the way on the right side, then Earth would be a quarter of an inch from the left side of the page—less than the width of your pinkie fingernail. Uranus would be about two finger-widths past the middle of the page. **Table 1.1** provides the distances for the eight planets in our solar system.

Tools of the Trade 1.1

Astronomers often deal with numbers that can be really large. For example, the distance from Earth to the Sun is about 150 million kilometers. Writing this number out would be cumbersome. It is generally easier to write them as powers of 10 instead, producing a more compact form. This format is called *scientific notation*.

In scientific notation, we write a number between 1 and 10 and multiply it by a power of 10. Thus, we can express 150,000,000 as 1.5×10^8, which is much easier to write and remember. Converting a number to and from scientific notation is done by moving the decimal point around, and is as easy as counting.

Scientific Notation

Converting to scientific notation:
1. Move the decimal point until it comes after the first nonzero digit.
2. Count the number of places the decimal point moved. This tells you the power of 10. It is positive if you moved the decimal to the left, and negative if you moved it to the right.

   Example 1: *The decimal point moves three places to the left.*
   $$2013 \longrightarrow 2.013 \longrightarrow 2.013 \times 10^3$$
   Example 2: *The decimal point moves three places to the right.*
   $$0.0042 \longrightarrow 0004.2 \longrightarrow 4.2 \times 10^{-3}$$

Converting from scientific notation:
1. The power of 10 tells you how many places to move the decimal point, either to the right (if it's positive) or to the left (if it's negative).
2. Fill in any spaces with zeros.

   Example 1: *The decimal point moves three places to the left.*
   $$4.2 \times 10^{-3} \longrightarrow .0042 \longrightarrow 0.0042$$
   Example 2: *The decimal point moves three places to the right.*
   $$2.013 \times 10^3 \longrightarrow 2013. \longrightarrow 2013$$

Astronomers are learning that the planets are only part of a solar system that is much larger than originally thought. Beyond Neptune exists an expansive ring of icy debris that we call the Kuiper Belt (rhymes with "viper"). Pluto, once considered the ninth planet of our solar system, is now categorized as a large Kuiper Belt Object and dwarf planet. This region may contain thousands or even millions of objects and is likely to be at least 10–20 AU in width.

But even this is not the edge of our solar system. Well beyond the Kuiper Belt is a region of space called the Oort Cloud (rhymes with "port"). This body of objects is believed to be generally spherical in shape and is thought to contain a huge number of icy objects left over from the formation of our solar system. The existence of both the Kuiper Belt and the Oort Cloud was deduced from comet orbits. We are now able to directly observe non-cometary Kuiper Belt Objects, and see infrared signatures of similar belts around nearby stars. Additionally, astronomers observe solar system comets with orbits that place their origin in the Oort Cloud.

Modeling indicates that if it does exist, it could extend 50,000 AU or more from the Sun. Referring back to our page-wide solar system scale model, if Neptune lies at the edge of our page then the Kuiper Belt would extend a fist-width off the right side of the page. The Oort cloud, on the same scale, would extend for three and a half football fields!

The nearest star, Proxima Centauri, is 4.24 *light-years* from Earth. This means that it takes a photon of light coming from this star 4.24 Earth years to travel from the star to your eye. If we scaled this distance down, once again comparing it to our page-width solar system, the nearest star beyond the Sun would be over a mile away!

What lies in the space between the nearest star and us? Essentially nothing. While there exists some trace amounts of gas and dust, for the most part the space between the stars is very empty.

Proxima Centauri is just the nearest star, however, and is not even bright enough in the sky to see with the unaided eye. The brightest star in our sky, Sirius, is 8.6 light-years away, which in our scaled-down model corresponds to roughly two miles away. The center of the Milky Way, our home *galaxy*, is actually some 20,000 light-years away from Earth and in our scaled-down model would be over 5,000 miles away from our textbook page! To consider the distances to other stars within the Milky Way, and the size of the Milky Way itself, we need to rescale our model solar system again.

The visible portion of the Milky Way consists of a flat, spiral-shaped disk surrounded by a spheroidal distribution of stars referred to as the halo. This disk is approximately 100,000 light-years in diameter. To begin to grasp the size scale of galaxies, let's scale down the Milky Way this time, so that its visible matter spans the width of this page. On this size scale, our solar system out to Neptune would be about 2 nanometers in diameter. A nanometer is one-billionth of a meter (10^{-9})—invisible to the eye and corresponding roughly to the size of a carbon atom.

> **Light-year**
> The distance that light travels through space in one year; 5.88 trillion miles (5.88×10^{12} mi).
>
> **Galaxy**
> A collection of billions to trillions of stars and star systems, all bound together by gravity.

Imagine one atom from the page of your textbook in this model, and that is how the solar system compares to the size of our Galaxy.

The nearest major galaxy outside of our own is the Andromeda Galaxy, depicted in **Figure 1.4**. This galaxy lies 2.5 million light-years away, which in our new scale model would be located over 16 feet away. The Andromeda Galaxy and the Milky Way Galaxy are just two galaxies among 50 or so bound together within a collection called the Local Group. This small conglomeration of galaxies is approximately 10 million light-years across, with its center being located somewhere between the Milky Way and Andromeda. Using our scale model, this would correspond to a diameter of 67 feet (a bit longer than a typical semi truck), with our textbook page Galaxy somewhere near the middle. Within the Local Group, our Milky Way is flying toward the Andromeda Galaxy and will one day collide and merge with it.

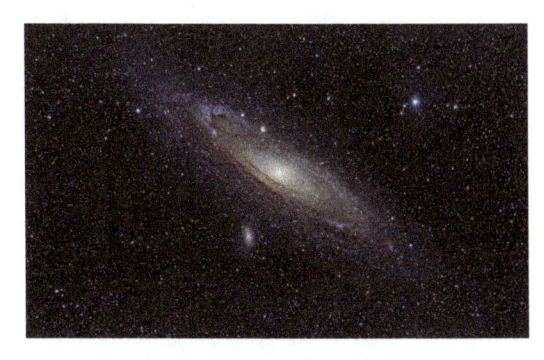

Figure 1.4. The Andromeda Galaxy

Copyright © Adam Evans (CC by 2.0) at http://commons.wikimedia.org/wiki/File:Andromeda_Galaxy_(with_h-alpha).jpg.

Groups of galaxies are the smaller cousins of much larger collections called galaxy clusters. While groups typically contain fewer than 100 galaxies, clusters can contain hundreds or even thousands of galaxies, as illustrated in Figure 1.5. The nearest big galaxy cluster is the Virgo Cluster, composed of roughly 2,000 galaxies. This massive structure is located about 60 million light-years away, putting it at a distance of 400 feet in our scale model. The Virgo Cluster and our Local Group are gravitationally bound together, along with other groups and clusters, to form a structure called the Virgo Supercluster.

From **Figure 1.5**, you can get the sense that typical galaxy separation distances are comparable to the sizes of the actual galaxies. On the other hand, typical stellar separation distances among stars are huge compared to the sizes of the stars. From this you can gather that stars are generally isolated form one another and rarely interact, while galaxies are often nearby other galaxies and almost inevitably interact.

Figure 1.5. A galaxy group called Stephan's Quintet (left) and the galaxy cluster Abell 1689 (right).

(left) NASA, ESA, and the Hubble SM4 ERO Team, "Stephan's Quintet," http://hubblesite.org/gallery/album/galaxy/cluster/pr2009025c/. Copyright in the Public Domain; (right) NASA, ESA, L. Bradley (JHU), R. Bouwens (UCSC), H. Ford (JHU), and G. Illingworth (UCSC), "Abell 1689," http://hubblesite.org/gallery/album/galaxy/cluster/pr2008008b/. Copyright in the Public Domain.

The most distant galaxies yet observed are found at such large distances that the light we receive is thought to have been emitted over 13 billion years ago. This corresponds to such astonishing distances that our brain begins to struggle to grasp them even with the use of analogies. In our scale model above, the most distant galaxy would be located over 15 miles away at the time its light was emitted. The expansion of the Universe has since dragged it even farther from us by now, making it even more distant. The Universe is truly an enormous place.

However, this is only the neighborhood of the Universe that we can see. Since the Universe has a finite age, light we observe has only had so much time to travel from the most distant locations. This means that we are effectively surrounded by a imaginary boundary, beyond which light has not had time to reach us yet. This defines our **observable universe**.

Figure 1.6. The Hubble eXtreme Deep Field, looking at galaxies in the field out to 13 billion light-years away.

NASA, ESA, G. Illingworth, D. Magee, and P. Oesch (University of California, Santa Cruz), R. Bouwens (Leiden University), and the HUDF09 Team, "Hubble eXtreme Deep Field," http://hubblesite.org/gallery/album/the_universe/pr2012037a/. Copyright in the Public Domain.

What is beyond this imaginary boundary? More Universe! It probably looks a lot like the portion of the Universe we see currently, and probably behaves in the same ways we are familiar with.

1.4 What does the Universe look like?

The Universe contains a myriad of different structures, from stars and planets to star clusters, galaxies, and galaxy clusters. Zooming in on the smallest size scales allows us to see the diversity of objects and structures that the Universe provides. Taking a step outward, astronomers see that galaxies and galaxy clusters are actually quite uniform in their distribution, as demonstrated by the Hubble eXtreme Deep Field (XDF), released in 2012. This image, shown in **Figure 1.6**, is the result of two million seconds of exposure and represents a look out across the Universe and back in time.

Indeed, on the largest size scales, the Universe appears remarkably smooth. Data collected by the Wilkinson Microwave Anisotropy Probe (WMAP) have shown us that the early Universe was smooth to within a factor of 1 in 5,000.

Figure 1.7 shows an all-sky map of microwave radiation believed to have been produced shortly after the Big Bang, referred to as the cosmic microwave background (CMB). The different colors correspond to different temperatures in the early Universe, where red is slightly hotter than average and blue is slightly cooler. While the image seems to show a Universe that is quite inhomogeneous, these temperature differences actually reflect a difference of just 0.0002 degrees K. On a ruler, one part in 5,000 is smaller than the eye can see. These temperature fluctuations, albeit present, are very small! Toward the end of this book we will discuss the importance of these temperature fluctuations in greater detail.

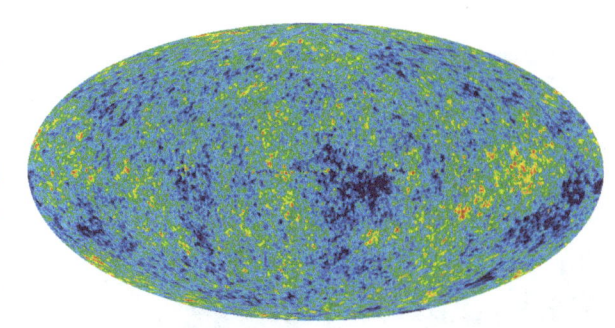

Figure 1.7. Results from WMAP show us that the Universe is remarkably smooth. This image shows temperature fluctuations so small that they had to be magnified thousands of times in order to be visible.

NASA / WMAP Science Team, "Nine Year Microwave Sky," http://map.gsfc.nasa.gov/media/121238/index.html. Copyright in the Public Domain.

Simulations of the evolution of the Universe using the most recent measurements of fundamental parameters produce a picture like that shown in **Figure 1.8** from the Millennium Run, a 3-D simulation that modeled the evolution of some 20 million galaxies over the course of about a month of computing time. This remarkable simulation, and subsequent runs by the same group of researchers, revealed that on the largest size scales the Universe is just as uniform as the CMB suggests, and that gravity from dark matter associated with all visible galaxies produces a filamentary structure—a kind of "cosmic web"—throughout the Universe. Galaxies and galaxy clusters all fall along these gravitational threads, leaving voids scattered throughout space where relatively little luminous matter appears to exist.

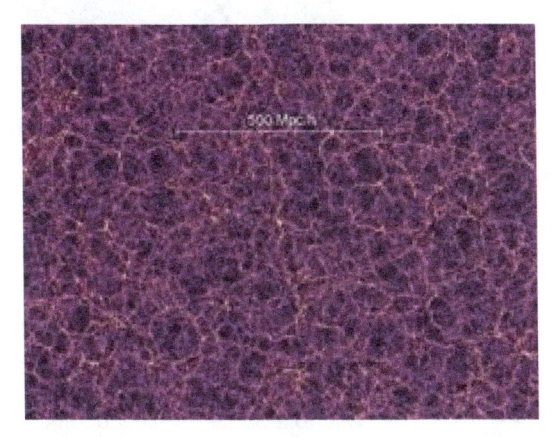

Figure 1.8. The Millennium Simulation, a computer model that traced the evolution of a cubic region of the Universe 2 billion light-years to a side.

Copyright © 2005 by Springel et al. / Max-Planck-Institute for Astrophysics

Figure 1.9. A slice of a 3-D map of the distribution of galaxies out to a distance of nearly half a billion light-years. The black strip down the middle represents the area of the sky partially obscured by dust in the Milky Way.

2MASS, T. H. Jarrett, J. Carpenter, and R. Hurt / Public Domain

Figure 1.10. An artist's rendition of the formation of a solar system around the star Beta Pictoris. It is believed that the formation of our Solar System may have looked very similar to this.

NASA, FUSE, and Lynette Cook / Public Domain

Observational data from survey telescopes support this result. The Sloan Digital Sky Survey (SDSS) and the Two Micron All Sky Survey (2MASS) are two such surveys, intended to obtain imagery and measurements for millions of stars and galaxies. **Figure 1.9** shows the distribution of these galaxies out to a distance of nearly half a billion light-years as seen over the entire sky. Comparing this image with **Figure 1.8**, you can see that the results from the Millennium Run do indeed accurately reflect the appearance of our Universe.

We live in a Universe that is diverse and non-uniform on small scales and quite smooth on the largest scales. The Universe we can observe is believed to contain approximately 100 billion galaxies, each containing on average about 100 billion stars. This means that within the observable Universe there exist roughly 10^{22} stars—that's a 1 followed by 22 zeros. Write that number out to appreciate the magnitude. If you could count one star per second, it would take you over 300 *trillion* years to count them all.

1.5 What is the age of the Universe?

It was once thought that the Universe was eternal and unchanging. However, as observations and instrumentation have both improved, a view has emerged that is vastly different. We now know that not only is the Universe changing over time, but it may not have always existed either.

Interestingly enough, the fact that our night sky is dark actually revealed something about the history of our Universe to astronomers dating back to the 1600s. The argument, popularized by Heinrich Wilhelm Olbers in the 19th century and now referred to as "Olbers' Paradox", goes like this: If the Universe is infinitely old and static, filled with an infinite number of stars, then in any direction an observer looks his or her sight line should end at the surface of a star. If you imagine a series of concentric shells surrounding the Earth, then the stars within the nearby shell would appear brighter than those in more distant shells. However, the more distant shells would be larger and so contain more stars. This additive effect cancels out the dimming

Figure 1.11. A portion of the circumstellar disk around the star β Pic, imaged by the Hubble Space Telescope. Small knots can be seen in the disk, possibly revealing the presence of rings viewed edge-on.

Paul Kalas (STScI) et al., WFPC2, HST, and NASA / Public Domain of a star's apparent brightness with increasing distance, so the light from all of these stars should add up to produce a "night" sky that is tremendously bright. The dark night sky was therefore interpreted to indicate that the Universe is finite in age or finite in extent, or both. We now know that the Universe is indeed finite in age, though its true extent has yet to be fully understood.

Measurements of radioactive elements in the Earth's rocks seem to suggest that the Earth has been in existence for approximately 4.5 billion years. We expect that the Earth and the Sun likely formed at approximately the same time, so we can estimate the Sun to also be between 4.5 and 5 billion years old. The early Solar System was likely an extremely chaotic and volatile place, with collisions between small objects occurring frequently, contributing to the growth and development of larger bodies. As these objects grew, they would have swept up the dust and debris left over, ultimately clearing out orbits and establishing a well-defined solar system.

Observations of such proto-solar systems around other stars strengthen our belief that these predictions are accurate. Circumstellar disks—rings of dust and debris surrounding other stars—have been detected throughout the local region of our Galaxy, indicating the early formation of planets. **Figure 1.10** depicts an artist's rendition of what this might look like around the star Beta Pictoris (β Pic), a star visible from the Earth's Southern Hemisphere. Such disks have been directly imaged, an example of which is shown in **Figure 1.11**, and the presence of substructure within these disks may suggest the presence of budding solar system objects.

> **Metals**
> A term used by astronomers to refer to any chemical element on the periodic table heavier than helium.
>
> **Globular Cluster**
> An ancient collection of hundreds of thousands of stars, all bound together by gravity into a spherical cluster roughly 100 light-years in diameter.

Figure 1.12. The globular cluster M80, which contains some one hundred thousand metal-poor stars and is believed to be approximately 12 billion years old.
AURA, STScI, and NASA / Public Domain

Heavy elements such as those used in estimating the age of the Earth are not expected to have been created in events that unfolded immediately following the Big Bang. Instead, heavy elements, referred to by astronomers as *metals*, are believed to have been produced through consecutive rounds of star formation, evolution, and death. The Sun contains a relatively high amount of such metals, meaning that it was not formed shortly after the Big Bang but came about following several generations of star formation. This implies that while the Sun may be 4.5 to 5 billion years old, the Universe itself may be much older.

Since it is expected that metal content within a star must correlate with the era during which that star formed, it would seem possible that some stars from the first generation, lacking any metals, might still exist. Star clusters called *globular clusters* contain hundreds of thousands of so-called metal-poor stars, with some clusters possessing stars with less than one one-hundredth the metal content of the Sun. **Figure 1.12** shows an image from the Hubble Space Telescope of the globular cluster Messier 80 (M80). Since the stars within a cluster are believed to have generally formed at nearly the same time (or within 100–200 million years of each other), clusters are helpful for testing our understanding of stellar evolution. By applying what is known about stars to these clusters, astronomers have determined that they may be as old as 12–14 billion years. This ranks them as some of the oldest structures in the Universe.

But is the Universe even older than this? Astronomers in the mid-20th century were faced with a conundrum: their estimates of the ages of the oldest stars actually exceeded some estimates for the age of the Universe. These estimates were based on the discovery in 1923 that the Universe is expanding. By turning the observed rate of expansion backwards, the age of the Universe could be estimated by determining how long ago the galaxies were all on top of each other. However, this initially ended up providing an age that was much lower than the ages of the star clusters. How could the Universe be younger than its contents?

Further research and observations helped resolve the discrepancy, and the latest results from the Planck satellite have provided the most precise estimate yet—as of March 2013, the Universe has been measured to be approximately 13.798 billion years old, with an uncertainty of about a quarter of a percent. Our best measurements now suggest that the first stars formed less than 1 billion years after the Big Bang. The determination of the estimated age of the Universe is described in greater detail throughout this book.

1.6 How confident are we in our knowledge?

With all of the claims that astronomers make, many of which have just been discussed, it's a reasonable question to ask: what makes astronomers so sure about these things? How can researchers take something so seemingly unfathomable as the Universe and determine its size, age, history, and future?

It is certainly true that intellectually honest scientists must at some point admit that our understanding about the Universe is limited by one fact: we will never be 100% certain of anything. It's impossible to measure or calculate any value exactly. However, while this is the case, we can make meaningful measurements and assertions with a high degree of confidence. We are by no means forced to throw up our hands in defeat.

> Uncertainty is a fact of life. We can never know something with 100% certainty, nor measure something with perfect accuracy.

The process of science involves first knowing what we can and cannot assert. In order to "do" science, we must know what science "is" and "is not."

First, what science is NOT:

- A process to solve all problems and questions.
- A process that can ignore rules.
- A process that attempts to prove things or establish absolute certainties.
- A process that can be relied upon for its complete objectivity.
- A process that is free from biases and opinions.

Does anything in this list of items surprise you? We often think of science as having at its core the drive to answer all questions with absolute certainty. While this is a noble desire, there are things that science cannot address directly, such as questions regarding the supernatural. The supernatural is, by definition, beyond the realm of the natural world and thus is not something to which we can apply rules and objective processes to study. We rely on science to tell us how the natural world generally works. This being said, science can neither prove nor disprove the existence of supernatural beings, nor can it rule out the possibility that supernatural beings might occasionally act in the world in ways outside of the ordinary laws of nature.

What about the third item on the list? As has already been mentioned, try as we might, there are no measurements that scientists can make that will be known with absolute certainty. There are always levels of uncertainty to each measurement and calculation. Various factors contribute to uncertainty, including ambiguous assumptions, faulty instruments or techniques, and poorly known quantities.

For example, suppose we were to try measuring the length of a table with a meter stick. We would lay the meter stick down, line up one end with the edge of the table, and read off the value on the meter stick that lines up with the opposite edge of the table. Perhaps we arrive at a length of 2.355 meters. Can we say with absolute certainty that the table is exactly 2.355 meters in length?

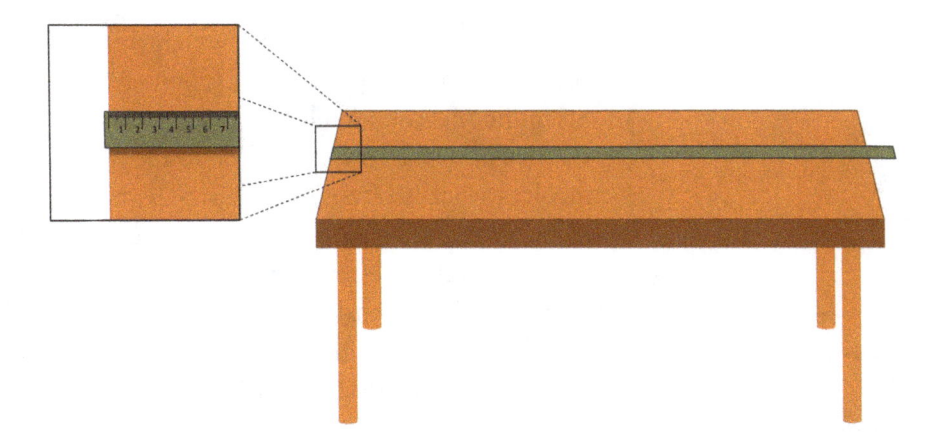

Figure 1.13. In measuring the length of a table, how certain are you that you have measured the table's true length? How certain can you be? What limitations might you experience, and what factors might play into your level of uncertainty in your measurement?

How well did we line up the end of the meter stick with the edge of the table? Is the end of the meter stick perfectly smooth, or is it beat up from repeated use? Does the meter stick's zero point coincide exactly with the end of the meter stick, or is there a small gap? How wide is each tick mark on the stick? Does the zero point of the stick start at the left edge of the first tick mark, or in its middle? Is the stick lying perfectly parallel with the table or is it angled at all? When we measured the additional 0.355m, did the edge of the table fall exactly at the 35.5 cm mark, or was it slightly before or beyond it? By how much? If we repeated the measurement, would we get the same number? What if we used a different meter stick? What if someone else took the measurement?

All of these questions must play into our determination of the table's true length. If this seems overwhelming to you, that's actually a good thing! Scientists must be extremely careful to assess every factor when they make measurements or calculations, because if any of these questions has an answer that brings doubt into the final value, then it must be considered as contributing to the uncertainty of the value. Sometimes the uncertainty is estimated, other times it is calculated. In the table length example, we might assess all sources of uncertainty and estimate that we could be off, at most, by 1 millimeter (0.001 m). We would then quote the table's length as being 2.355 ± 0.001 m.

So for all that science isn't, what *is* science? Science can be summarized as:

- A rules-based process of observation and testing.
- A process that attempts to generate new observations and data which will be either consistent or inconsistent with theories.
- Prone to human error or interpretation bias.
- Uses peer review to reduce, but never fully eliminate, errors and bias.
- A process of eliminating explanations until the one left standing is the "best solution."

> **Science**
> A process of observing the world around us and testing possible explanations as to its behavior until we arrive at an explanation that consistently and accurately describes the observation.

> **Hypothesis**
> A tentative explanation for an observation that provides testable predictions.

So what is the process by which this is done? The procedure that is typically used follows what is called the Scientific Method. While the exact procedure can often vary slightly from one discipline to the next, it typically involves what is often a cyclic process of idea development, testing, and adjustment, illustrated in **Figure 1.14**.

In scientific methods, we start with an observation regarding the natural world around us. We might ask ourselves why something ought to appear the way it does, or what caused an event to occur as it did. This gets the process rolling.

In order to approach our question scientifically, we generally then come up with a possible explanation: a

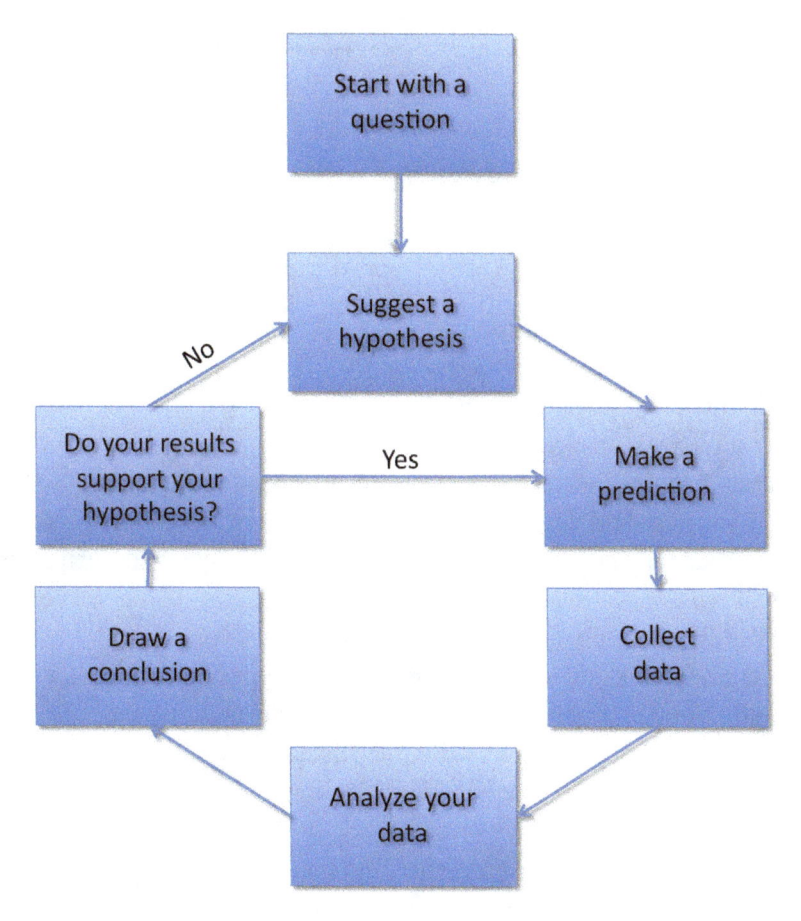

Figure 1.14. An illustration of a typical Scientific Method (simplified).

hypothesis. Using this hypothesis, we then come with a prediction that would support our hypothesis if correct. The aspect of making testable predictions is a hallmark of a scientific hypothesis; if we are unable to test an idea, then that idea must ultimately be relegated to the realm of philosophy. It is important to note that this does not make the hypothesis a bad idea—and indeed it might be useful in helping us think about the natural world and generate other ideas—it's just not scientifically testable.

Once we have our prediction in hand, we must test it somehow. This might mean setting up an experiment, or it might simply mean going out and making additional observations. In chemistry, experiments are carefully designed to test predictions. In astronomy, however, we have no control over our research subjects (stars, planets, galaxies, etc.) so all we can do is make observations. If our observations do not support the hypothesis we made—perhaps they contradict our prediction—then we must modify our hypothesis or throw it out altogether and develop a new one.

Fact: We perform the scientific method all the time, every day!

If our observed results agree with the prediction and support our hypothesis, then we can either refine our experiment for greater precision, perform additional tests or make even more observations, or develop another prediction that stems from the hypothesis and test that one next.

> What other examples from your daily activities can you think of that illustrate how you applied the scientific method, perhaps without even knowing it? In the days ahead, be aware of the observations, predictions, tests, and conclusions you make and see how the scientific method is actually a very familiar experience.

Little do you know it, but you actually live out this method in one way or another every day. Life is full of observations and many of them require us to make predictions and determine whether or not we were right. For example, suppose you were driving down the road in your car and suddenly you saw droplets of water hit the windshield. You might hypothesize that this means it is raining outside, and predict that if you turn on your wipers then it will wipe away the water as it continues falling. You test this prediction by turning on your windshield wipers.

But what if you notice that after a couple wipes there appear no additional water droplets on your windshield? Further observation reveals that the sky is sunny, with no clouds in sight. This would suggest that perhaps it wasn't raining. Instead, you might revise your hypothesis to involve the car in front of you washing its windshield, with some of their washer fluid overshooting their car and hitting yours.

You would test this by making an observation: do you see their wipers moving? Does their windshield look freshly wet and wiped? The observations you make will ultimately tell you whether or not that hypothesis was correct.

This is a simple example, but powerful nonetheless. The process of science, and science itself, is something that all of us do on a regular basis. This makes it one of the most familiar experiences you will have in life. Many people feel intimidated by science because it feels very foreign to them. In some cases, the subject may be unfamiliar (we don't often find ourselves studying wavelengths of light as part of our daily routine), but the process itself is no different than how we go about our lives. Life is full of trial and error, and a careful inspection of **Figure 1.14** will reveal that this describes the scientific method quite well.

Not all hypotheses remain hypotheses, though. If a hypothesis is repeatedly, and without fail, supported by a myriad of tests then it may eventually become known as a *theory*. Theories are generally simple, elegant ways of describing nature around us that are always supported by experiments and observations.

In some cases, if a theory is upheld for long enough under significant testing, it may become a scientific law. A primary difference between a theory and a law, however, is that laws are generally descriptive of some physical phenomenon, whereas theories often provide an explanation for the phenomenon. Newton's Law of Gravitation, for example, describes how gravity is manifested, whereas the General Theory of Relativity describes why. Scientific models are constructed on the basis of well-understood physical principles and highly-tested theories, and can serve as powerful tools for explaining observations or making additional predictions for further study. In this book, when we speak of scientific models we will be referring to them with a high degree of regard, as the culmination of the best application of the scientific method to date.

In everyday terms, the word "theory" is often used where the word "hypothesis" should be used instead. Thus, when people refer to the "Theory of Relativity," for example, it is sometimes done in such a way as to dismiss or debase the model—in many cases following up with the phrase "It could be wrong. It's just a theory."

This usage actually matches more closely with a dated use of the terms *theory* and *law*. In the past, the term *theory* was applied to what we would now call a hypothesis. In that historical language, a *law* corresponded to what we now call a theory. In the modern era, these terms have become more refined. This is partly why there still exists some confusion when the terms are used, and why a certain amount of scientific literacy can be helpful.

While we've seen that it is scientific practice to accept that nothing can be proven beyond a shadow of a doubt, valid scientific *theories* are given such a title only when they have been upheld time and again. If at any time the predictions of a theory fail to match an observation, then the theory must be discarded or modified to accommodate the new observation. Valid theories are grounded in an overwhelming body of evidence.

> **Theory**
> A simple, powerful hypothesis for describing the natural world that is consistently supported by a variety of tests.

This being said, there are some cases where the word "theory" is being misused in modern language, even in the scientific world. "M-theory," an idea that will appear later in this book, is such an example. This idea, while founded on verifiable mathematics, does not at this point make many testable predictions that allow scientists to distinguish between different versions of the idea or contrast the predictions and results with other possible explanations. It is possible that as this idea develops further it may become more testable and eventually supplant our current model, but for now the term "theory" that is used to designate it is technically a misnomer according to modern usage as described here. It should, perhaps, be called "M-idea".

As you read through this book, it is your role to not only try to understand the concepts but to evaluate them as well. With each topic and chapter, critically analyze the ideas being presented and decide for yourself if the explanations make sense. Does the evidence give adequate support for the explanation? Do the explanations have weaknesses? Feel free to consult other legitimate, scientific resources to see if additional observations exist that add further credibility to the explanations. Alternatively, consult your instructor—they are happy to help. In many instances, researched critical analysis like this has led to fresh insight and new ideas.

CHAPTER TWO

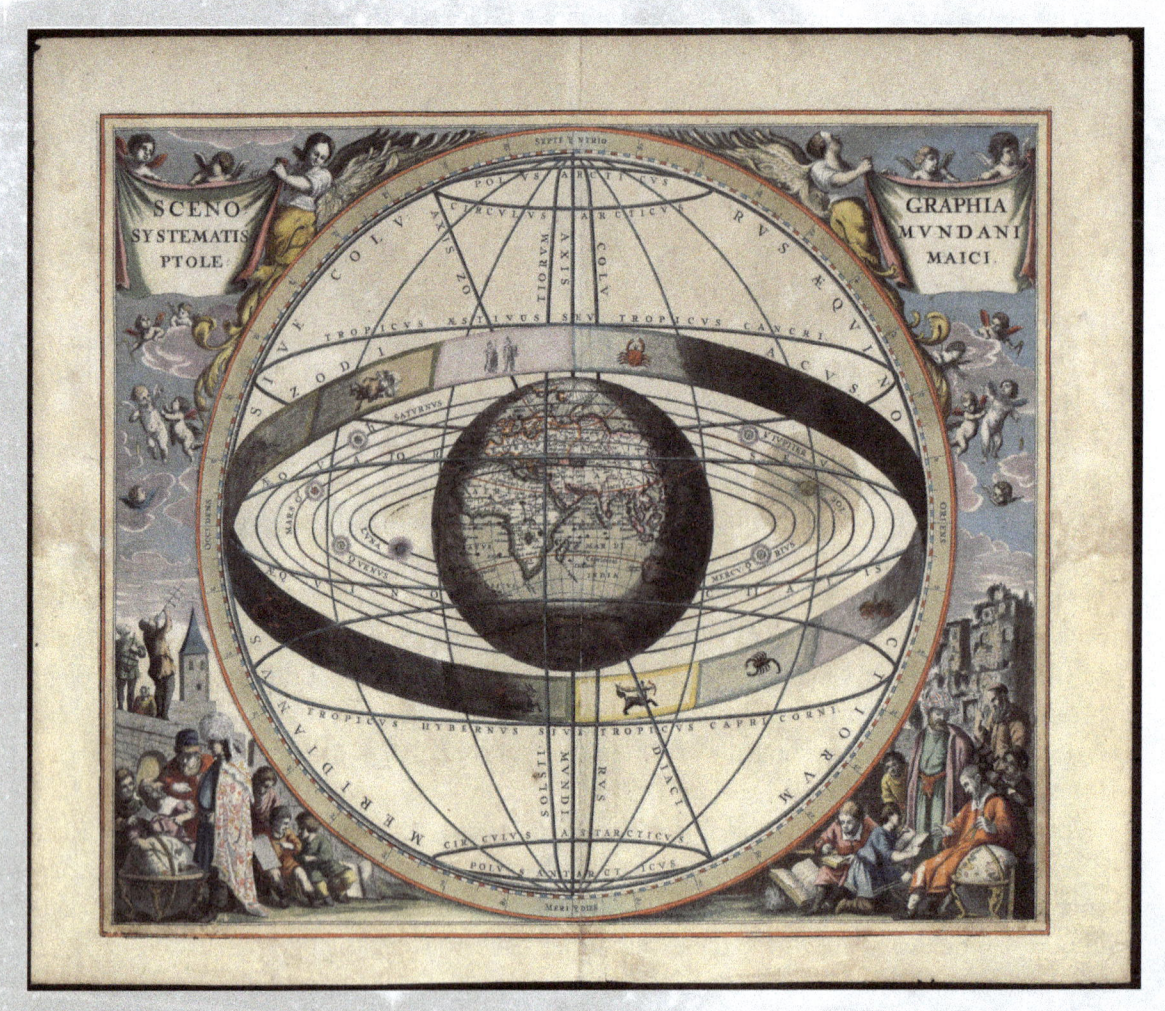

Chart painted by Johannes van Loon (1660) depicting the Ptolemaic geocentric model of the Universe.
Johannes van Loon / National Library of Australia / Public Domain

Ancient Ideas

Learning Objectives

In this chapter you will learn:
- » What ancient civilizations believed about the creation of the world and humanity
- » What factors impacted the development of ancient creation accounts
- » How ancient civilizations viewed and studied the sky
- » Why the ancient Greeks concluded that Earth is at the center of the Universe
- » How the ancient Greeks explained the motion of the celestial objects

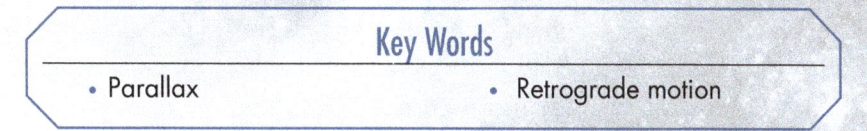

Key Words
- Parallax
- Retrograde motion

2.1 Cosmology of ancient civilizations

At the fundamental level of human experience, one question is continually present: Why am I here? It's common and natural for us to pursue such introspection as individuals. As collections of individuals, entire civilizations have asked the same question. Why are we here? Where did we come from? Do we have a purpose? Where did all that we see come from? How was it made? Other typical questions might have included: What is my place in society? What is our relationship with or obligation to the world around us?

As answers to these weighty questions were sought and developed, religious systems generally followed. These systems often involved pantheons of gods and goddesses and began with stories (myths) describing the creation of the Universe, the Earth, and humanity. The word 'myth' derives from the Greek word *mythos*, which refers to a story or set of stories that were

Figure 2.1. Quetzalcóatl, as depicted in the Codex Magliabechiano (16th century).
Wikimedia Commons / Public Domain

passed on verbally from one generation to the next and that had a significant meaning for a particular people group. Each myth contained details specific to the individual culture, but perhaps surprisingly there were many elements and characteristics that were common to these stories overall. While all of these stories may or may not have been intended to represent a culture's true conception of how creation occurred, they all are valuable for what they have to say about relationships and values.

Let's look at a few examples from ancient civilizations from around the world. These examples are but one version attributed to each culture, and versions may vary slightly by regional recollection or translation. Nevertheless, as you read the following paragraphs, see if you can identify elements unique to the culture as well as overarching concepts that they share. Each of these accounts varies richly from the rest as a reflection of the culture from which it originated and the environment in which those people lived.

Ancient Americas

The ancient ancestors of the Toltec people of what is now Mexico, predecessors of the Aztecs, spoke of a singular supreme god named Ometéotl, who was considered a dual god—that is, he was both father and mother of all things. He caused the birth of four sons collectively called the Tezcatlipocas who were identified as the primordial forces of creation. Through them, space and time came about. After a series of cataclysms where each brother attempted to assert his dominance over the others by creating an Earth with various inhabitants, only to be met with destruction at the hands of the other brothers, the four of them ultimately worked together to reestablish the Earth peacefully. Two of them became the Sun and Moon, while one, Quetzalcóatl, illustrated in **Figure 2.1**, created a man and a woman. His blood gave them life and corn gave them sustenance.

The Mayans, in a collection of stories called the Popol Vuh, described a series of attempts at creating mankind to be servants and worshippers before finally succeeding. The first population of humans were made from mud, but dissolved after soaking up water. The next humans were carved out of wood, but lacked souls and were destroyed by a flood. Finally, a man and a woman were created out of cornmeal and the human race sprang from them.

Pacific Islands

Polynesian mythology envisioned all of creation as a coconut shell, containing various lands within it. The regions in the stem of the coconut were where various spirits lived that sustained and gave life to the world. In its base lived a woman named Vari-ma-te-takere who began plucking off bits and pieces of her flesh to form gods and humans. The first human, Vatea was the father of all subsequent gods and men and was half man, half fish. In some renditions, his eyes represent the Sun and the Moon. He had a brother named Tinirau, also half man, half fish, who became the god of the sea creatures. Numerous other progeny were created who became gods and goddesses of animals and weather patterns. Each land within the coconut was inhabited by various descendents of Vari-ma-te-takere's progeny, with humans living at the top in a place called Te-papa-rairai, or The-thin-land.

Asia

The civilization with perhaps the greatest variety of creation stories is that of the ancient Chinese. While the details of the mythology seem to change every few hundred years, a common theme of the balance between yin and yang persists throughout. In the earliest recorded stories, there was a singular force in place in the beginning called Tao. The Tao set the creative events into motion, first forming unity, which then produced the duality of Yin and Yang. This duality, a fine balance, ultimately led to the creation of all the complexity of the Universe, including people.

Figure 2.2. A 15th-century artist's illustration of P'an Ku.

Wang Qi / Public Domain

In later retellings, sky, earth, and light formed out of chaos and combined with the essences of yin and yang to form a man. This man, named Huang-lao, was the first human and was taught many things by a golden being from the heavens who appeared to him.

A hero god-giant named P'an Ku (or Pangu), depicted in **Figure 2.2**, appears in many roles throughout the variety of Chinese mythological stories. In some versions, a chicken egg is formed out of the primordial chaos. This egg was in essence the commingling of heaven and Earth. When the forces of yin and yang had attained balance, P'an Ku emerged from the egg by splitting it into two pieces—thus separating heaven and Earth—and sculpted the Earth while also creating the Sun, Moon, and stars. Following these creative acts, some versions have him creating and teaching humanity while other versions depict his death as the final act of creation—with all aspects of his body producing such things as soil, plants, stones, rain, and (from the parasites inhabiting his body) people. Interestingly, several interpretations of the P'an Ku story exist, along with many different understandings about its true origin within the Chinese culture or from elsewhere.

Africa

Egyptian creation mythology also changed as the civilization evolved. Changing dynasties often brought about new emphases on different gods (and even the introduction of new gods), which necessitated the modification of mythology as well. The earliest accounts tell of a lifeless chaos of water, out of which rose a pyramid-shaped mound. Out of this mound came the god Ra, though various versions refer to him as Ptah, Atum, or Amun. He produced, among other deities, the air god Shu and goddess of moisture Tefnut. They gave birth to the sky goddess named Nut and an earth god named Geb. Shu had to separate Nut and Geb, lest they procreate, and thus the Earth was formed, with Egypt and the Nile River in the center. Some versions say Ra then populated the Earth with humans formed from his own tears, while others say humans were molded from clay by the god Khnum, who is generally associated with creation and the Nile River. **Figure 2.3** shows an illustration of these Egyptian deities.

Figure 2.3. Geb, god of the earth, reclines at the bottom of this painting while Shu, god of the air, holds up Nut, goddess of the sky. Also shown are other deities traveling through the sky in their boats.

E.A. Wallis Budge / Public Domain

Europe

While fewer written records exist from ancient European civilizations, the ancient Greek mythological traditions, which were subsequently adopted by the ancient Romans, have been well documented. As the culture changed in response to the introduction of outside influences, the mythology adapted with it. Ultimately, this means that any story attributable to the Greeks may not have actually been told throughout the entire history of the civilization.

In one popular rendition, the world began as Chaos and out of this arose Gaia, who without the help of a man gave birth to a male child named Uranus. Gaia and Uranus together then had a number of offspring known as the Titans. The Titans were immortal and ruled early on as gods and goddesses, only eventually to be overthrown by Zeus, the son of one of the Titans. Zeus then took over as king and became the father of a new generation of gods called the Olympians. Prometheus and Epimetheus, two of the Titans who did not participate in the battle and were thus allowed to live, set about creating men out of clay.

Figure 2.4. 1802 painting by Jean-Simon Berthélemy depicting the creation of man by Prometheus, in the presence of Athena.

Jean-Simon Berthélemy & Jean-Baptiste Mauzaisse / Public Domain. Photo by Marie-Lan Nguyen.

Prometheus was very happy with the men he had created, but when Zeus decreed that men must present sacrifices to the gods, Prometheus helped the men deceive Zeus with a sacrifice of bones wrapped in fat. One of the punishments for this deception was that Zeus had a woman created named Pandora, the first human woman, who was given great beauty but also a lying tongue and deceptive heart. She was sent to the men with a mysterious container (in some accounts a box, in others a jar) that she was not allowed to open. Letting her curiosity get the better of her, she opened the container and released all of the evil and sorrow into the world.

Middle East

Interestingly, creation accounts from ancient civilizations in the Middle East share many common elements. For example, the Judeo-Christian account of Genesis, the Mesopotamian Epic of Gilgamesh, and several other Sumerian and Akkadian stories all tell of a garden or city where creation began.

In the Judeo-Christian account of Genesis, one eternally existing god named Yahweh created the universe, formed Earth, and designed all of its creatures in a six-day period. This single, eternally existing god marks a distinction from the other stories where one or more gods are born or created. His final creations were Adam, the first man, and Eve, the first woman. Adam and Eve were originally intended to live forever in harmony and close relationship with Yahweh, rather than the enslavement that was common in other cultural stories. However, by giving in to temptation they defied Yahweh and were subsequently cast out of the idyllic Garden of Eden into the world.

Details of an ancient flood, also described by the Mayans and many other Old and New World cultures, are given in both the Judeo-Christian book of Genesis and several other stories of Mesopotamian origin, including the Epic of Gilgamesh. While the recounting of the flood is generally cast as a punishment on the world by one or more gods, the common elements of this story shared by a variety of ancient civilizations have led many scholars to believe it to be drawn from an actual historic event. However, this idea is still widely debated.

Figure 2.5. A depiction of Noah's Ark and the Great Flood, by American folk artist Edward Hicks.

Edward Hicks / Philadelphia Museum of Art / Public Domain

2.2 Ancient sky watchers

While most, if not all, civilizations developed some collection of stories that served to explain the origins of the world and humanity, many ancient peoples were also accomplished observers of the night sky. The ancient Chinese were exceedingly meticulous about recording positions of stars in the sky, even creating maps of their own constellations (**Figure 2.6**). Through careful observations, they were able to predict the occurrence of lunar eclipses to great accuracy and made records of solar eclipses, comet appearances, and supernovae.

The ancient Egyptians were another civilization that carefully kept track of the stars. The annual flooding of the Nile River, which brought with it both nutrient-rich silt for farming and destructive high-water levels,

Figure 2.6. A section from the Dunhuang Star Atlas, dating to around 700 AD, which provides detailed maps of ancient Chinese constellations. In the one shown here, the asterism we now call the Big Dipper can be seen along the bottom.

Wikimedia Commons / Public Domain

Figure 2.7. The ancient Mayan observatory at Chichen Itza, Mexico.

necessitated the ability to predict its occurrence. By observing the yearly appearances of the stars, they were able to develop an accurate calendar for keeping track of the floods and seasons.

In Central America, the Mayans were also avid stargazers. Astute observations of the heavens allowed them to predict the occurrence of solar and lunar eclipses, the periodic appearance and disappearance of Venus, and the passing of the seasons. They even went so far as to build their own observatory, shown in **Figure 2.7**, and ultimately were able to make such precise measurements that they could derive the cyclical relationships between the planets, Sun, and Moon.

2.3 From supernaturalism to natural philosophy

Throughout the early millennia of human civilization, one common belief was generally held: we would never be able to truly understand the world around us. While we might be able to pick up on patterns and make observations, very few ancient cultures believed we could ever grasp the underlying principles and causes of what was observed. This is reflected in many of the ancient mythological stories as cataclysms coming forth based on conflicts and whims of the gods. Such belief ultimately led the ancients to marvel at unpredictable elements of nature—violent storms, earthquakes, plagues, etc.—while never truly seeking to understand why these events occur. And for that matter, why they don't always occur.

Around the 7th century BC, the ancient Greek civilization began toying with a new concept. Instead of dismissing the world as unknowable and remaining ignorant of its workings, several Greek thinkers began to ponder why the unpredictable events were so rare. For them to be rare, this meant that there was some version of normality that was more often experienced. They ultimately concluded that, in general, the world around them functioned in a very orderly way.

But why should this be? Furthermore, if the world is orderly, then it must function in ways that make sense. That is to say, there must exist a way of grasping how nature operates. This in and of itself does not dismiss with the notion of the supernatural, but rather insists that whatever "rules" have been laid out for the world to follow, these rules are comprehensible. For the ancient Greeks, accepting the notions that the world operates in an orderly way and that nature is comprehensible were implicitly acts of faith, since there was no definitive evidence to compel this acknowledgement. Modern scientists today still act on this basis. But if the potential to understand the laws of nature exists, then it became essential to study, explore, experiment with, and observe the natural world to develop this understanding.

And so began the practice of philosophy. Ancient Greek philosophers embarked on a journey of debate and discovery as they posed and considered elementary questions with fresh eyes. What is the fundamental nature of matter? Why do mathematical relationships exist? Does one explain the other? Greek philosophers examined a myriad of competing ideas and carefully ruled any out as they were able. In doing so, they became the world's first scientists.

2.4 Early thinkers

Ancient Greece and the empire of Alexander the Great was a hotbed of philosophical thought in the centuries from 600 to 100 BC. With their newfound freedom of thinking, new ideas came

Figure 2.8. Thales
Ernst Wallis et al. / Public Domain

Figure 2.9. Democritus
Public Domain

Figure 2.10. Pythagoras
Copyright © 2011 by Marie-Lan Nguyen / Wikimedia Commons / CC BY 2.5.

about. To address the topic of the fundamental composition of matter, Thales (ca. 624–ca. 546 BC; shown in **Figure 2.8**) proposed that everything was made of water in various forms. Realizing that water is one of the Earth's most abundant resources, it did not seem a stretch to wonder if everything was composed of this substance. As the idea was debated, a new model developed by Empedocles (ca. 490–ca. 430 BC) that listed four fundamental elements of nature: water, air, earth, and fire. Perhaps everything in existence could be understood as being some form of one of these elements.

Others, such as Leucippus (early 5th century BC) and his pupil Democritus (ca. 460–ca. 370 BC; shown in **Figure 2.9**), proposed that all matter might actually be composed of the same basic units, which are then simply arranged in different ways. This suggested that matter itself could not be subdivided into smaller units without end but that there exists some fundamental particle, which was called an "atom," that makes up everything in the Universe. This idea was remarkably prescient.

While many philosophers were speculating about the nature of matter, others were developing the subject of mathematics. Interesting and insightful relationships between numbers were being understood, and quantitative descriptions of shapes were being uncovered. Pythagoras (ca. 570–ca. 495 BC; shown in **Figure 2.10**) led the way in this pursuit, demonstrating his eponymous relationship between the side lengths of a right triangle. His adherents also were among the first to correctly suggest that the Earth is spherical by observing and explaining why certain constellations were visible from northern latitudes but not visible farther south. The Pythagorean school of thought would ultimately produce two of the greatest philosophers of ancient history: Plato and Aristotle.

Plato (ca. 428–ca. 347 BC; shown in **Figure 2.11**), taught by Socrates, was a prolific writer and teacher. His school of thought evolved into one that emphasized the importance of mathematics, geometry in particular, in our ability to understanding the Universe. As the circle was considered to be a perfect and pure shape, circular motion was generally adopted to reflect the purest motion. Spheres were the ultimate epitome of perfection and purity, and thus it was assumed that the celestial objects—Sun, Moon, Mercury, Venus, Mars, Jupiter, and Saturn—were spherical.

Figure 2.11. Plato (left) and Aristotle (right) are shown in discussion in The School of Athens, painted by Raphael.

Raffaello Sanzio / Public Domain

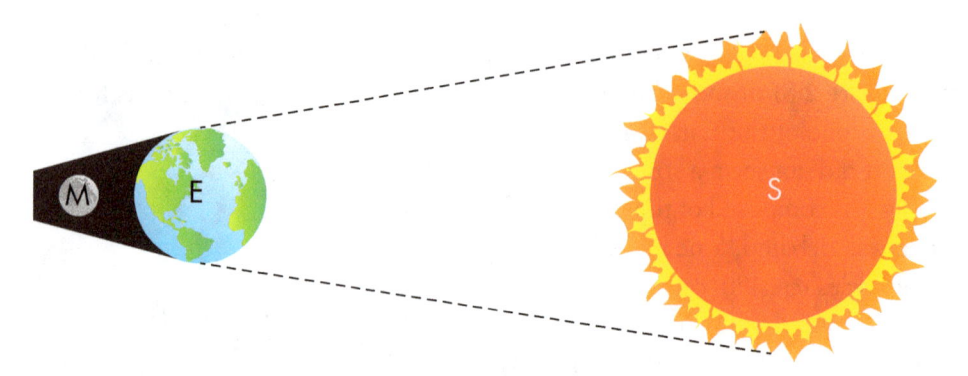

Figure 2.12. By noticing when the Moon (M) passed through the Earth's (E) shadow, Greek philosophers were able to work out its relative size compared to the Earth. Seeing that the Moon and Sun (S) had the same angular size, they used diagrams like these to work out the relative sizes of the Earth, Moon, and Sun. Sizes and distances in this diagram are not to scale.

Plato's protégé, Aristotle (384–322 BC; shown in **Figure 2.11**), initiated the merging of empirical observation into philosophical thought. While Plato had often approached subjects and questions in a deductive way (starting from general principles and coming to specific conclusions), Aristotle employed astute observations of the natural world and drew his conclusions inductively (starting from specific examples and developing general principles). Through his studies of biology and earth science, Aristotle gradually developed principles that became the foundation of the scientific method.

In his book *De caelo* (meaning "On the Heavens"), Aristotle described his observations of the stars during his travels, noting that *"there are some stars seen in Egypt and in the neighborhood of Cyprus which are not seen in the northerly regions,"* suggesting that our viewpoint on a spherical surface was the only possible explanation for this. He went on to note that *"stars, which in the north are never beyond the range of observation, in those regions rise and set,"* meaning that the observed patterns and motions of stars in the sky differ as one moves north and south. He went on later to observe that the Earth must not be terribly large, or such an observation would not be easily made. While it had been taught by some (including Plato) that the Earth was spherical, Aristotle was among the first to justify the claim with observational evidence and reasoning.

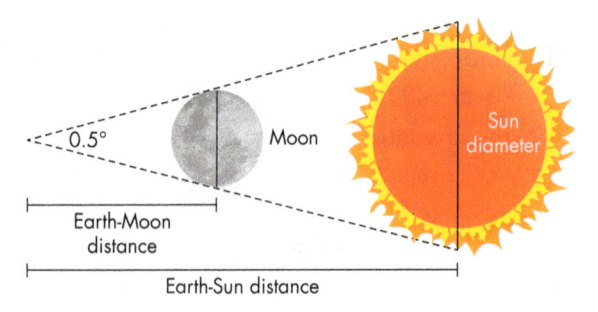

Figure 2.13. The Moon and the Sun have nearly the same angular sizes as viewed from Earth, so during a typical solar eclipse the Moon completely blocks out the Sun. Using this observation, it was possible to attempt to determine relative size and distance relationships between the Moon and Sun. Sizes and distances in this diagram are not to scale.

When it had been believed that the Earth was flat, it was clear that a person's weight held them on the ground. But with a spherical Earth, in which direction did that weight tend? Everyone knew that people in the southern hemisphere didn't simply fall off the Earth, so it was posited that perhaps everything in the Universe tended toward the "center" of the Universe. Evidently, then, this center coincided with the center of the Earth, and thus everyone on Earth was held in place. In addressing this question, the Greeks developed what became known as the geocentric ("Earth-centered") model of the Universe.

Before the age of philosophers, it had been assumed that the Sun and Moon were two deities in the sky, or at least were objects affiliated with deities. Any light they gave off was presumed to be intrinsic. However, in the 5th century BC, a philosopher named Anaxagoras made the rather audacious claim that the Moon did not glow of its own accord but instead simply reflected the light produced by the Sun. This claim came about after noticing that eclipses of the Moon occurred only when the Moon and the Sun were on opposite sides of the Earth. Since the Sun clearly gave off a tremendous amount of light and was responsible for lighting up the Earth during the day, it was concluded that the Earth probably casts a shadow. This shadow would point away from the Sun, so when the Moon passed behind the Earth then it would in some instances pass through the Earth's shadow, causing it to go dark (**Figure 2.12**).

By timing the Moon's passage through the Earth's shadow, and assuming that the Moon was much closer to the Earth than was the Sun, the Greeks were able to estimate that the Moon was approximately one-quarter the Earth's size. Given, then, that the Moon and the Sun had the same angular size—that is, the same apparent size in the sky—an observation easily seen during solar eclipses, they attempted to work out geometrically the size of the Sun compared to the Earth (**Figure 2.13**). However, in order to do so, they needed a little help to solve the puzzle.

Figure 2.14. Recognizing that the Moon's first-quarter phase occurred when the Sun, Moon, and Earth were at a 90-degree angle, Aristarchus was able to use the observed apparent angular separation between the Sun and Moon at that moment and determine the relative distances in the triangle shown above. Sizes and distances in this diagram are not to scale.

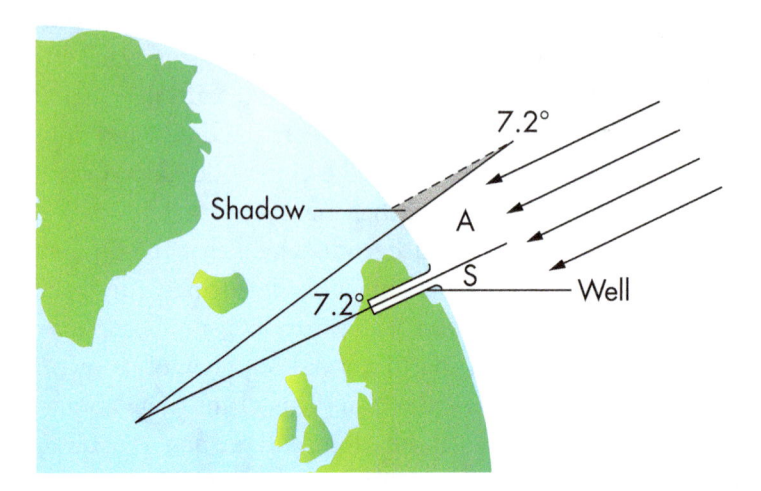

Figure 2.15. Parallel rays of sunlight that go straight down a well in Syene (location S) cast a small shadow for a vertical object in Alexandria (location A). The angle of this shadow is equal to the angle between A and S on the surface of the sphere (Earth). Knowing the physical separation between A and S can thus provide a measure for the full circumference of the sphere.

Aristarchus (310–230 BC) of Samos was a philosopher and mathematician who set about devising a way of determining the relative distances of the Sun and the Moon. To do this, he recognized that he could use geometry to create a triangle between the Earth, Moon, and Sun when the Moon was exactly at its first-quarter phase, shown in **Figure 2.14**. At this exact moment, the angle between the Earth and Sun, viewed from the Moon, was 90 degrees. Aristarchus then measured the angle between the Moon and Sun, viewed from Earth, and estimated it to be approximately 87 degrees. By applying principles of geometry, he was able to arrive at the result that the Sun was approximately 20 times farther from Earth than was the Moon. While history reports that his measured angle was off by a couple degrees, producing a Sun distance that was still low by another factor of 20, the fact that he was able to apply mathematical concepts to measure celestial parameters is a testimony to his intellect.

Aristarchus verified from these calculations that the Sun was quite distant compared to the Moon. Since they both have the same angular size (apparent size in the sky), this meant that the Sun is also quite large compared to both the Moon and the Earth—again, larger than the Moon by a factor of roughly 20. Some scholars believe it was this calculation that ultimately led him to conclude that the Universe was more likely to be heliocentric (Sun-centered) than geocentric, for why should the Sun, being so large, orbit the Earth when it would make more sense to place the largest object in the center instead?

However, this ran against what had become common knowledge. The geocentric model had been the favored view of Plato and Aristotle and was held in very high regard. Experience told the people of the day that the Earth "obviously" wasn't moving, for reasons that will be discussed in greater detail in the following section, and so the heliocentric model of Aristarchus was largely dismissed.

Eratosthenes (276–195 BC), the chief librarian at Alexandria in northern Egypt, was the first to turn the relative quantities of Aristarchus into true numerical values with dimensions. During his time at the library, he learned of a well in the town of Syene, located in southern Egypt, where the Sun shone all the way to the bottom at local noon on the summer solstice. He realized that in Alexandria, vertical objects cast small shadows at noon on the same day and that the angle of the shadow could be used to reveal the physical size of the Earth.

If the Sun is at a sufficiently large distance, then rays of sunlight are traveling in parallel lines by the time they reach the Earth. If one of these rays travels straight down to the bottom of a well, then the angle of a shadow cast by a vertical stick set up at a separate location on the Earth also represents the angle of separation on the Earth between the well and the stick. In geometric terms, these are similar triangles, as shown in **Figure 2.15**.

Eratosthenes knew that this separation angle, measured using the stick's shadow to be 7.2 degrees, represented the difference in latitude between Syene and Alexandria. Mathematically, the ratio of this angle to a full 360-degree circle was equal to the ratio of the distance between the two cities and the full circumference of the Earth. This separation distance, measured in Egyptian units to be about 5,000 stadia (approximately 785 km), when inserted into the relationship

$$\frac{7.2°}{360°} = \frac{785 km}{C}$$

produced a circumference measurement C of 39,250 km, a value accurate to within 2% of the actual value. This remarkable achievement demonstrated that all anyone needed was their mind and a tool and the physical world suddenly became measureable!

Once the Earth's size was known, the true physical size of the Moon quickly followed because the relative sizes of the Moon and Earth had previously been estimated from lunar eclipse observations. Additionally, since the Earth–Moon and Earth–Sun distances had been determined in units of Earth radii (**Figures 2.13** and **2.16**), it was a quick calculation to figure out those distances in physical units. Finally, the Sun's physical size was estimated using the comparable angular sizes of the Moon and Sun, and a little geometry (recall **Figure 2.13**). However, errors in measurement of the Moon–Earth–Sun angle (shown in **Figure 2.14**) produced a large degree of uncertainty in the Earth–Sun distance. Obtaining a more accurate estimate would have to wait almost two thousand years.

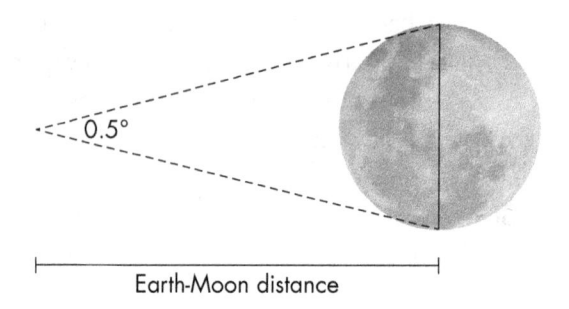

Figure 2.16. Using the Moon's apparent angular size of half a degree, combined with the estimate of the Moon's diameter equal to roughly one-fourth that of the Earth (see Sec. 2.4), trigonometry can reveal the Earth–Moon distance in units of Earth radii. This was then used to represent the Earth–Sun distance in units of Earth radii as well (Fig. 2.13).

2.5 The geocentric universe

Ancient observers were well aware that there were additional moving objects in the sky besides the Sun and Moon. These dimmer points of light, named 'planetes' in Greek (meaning "wanderers"), moved across the sky against the fixed background of stars. It had been noticed that these motions were periodic in similar ways to the motions of the Moon and Sun, and it was therefore assumed that these objects—again originally attributed to the deities—also followed circular paths around the Earth.

While it may be taken for granted in our modern era that the Earth orbits the Sun, this notion in ancient times was not a given, and for three very practical reasons:

1. **If the Earth was moving, we should feel a wind as we move through space.** Our human senses provide very effective insight into the world around us through the way we interact with it. In common experience, whenever a person is in motion, he or she can certainly detect this motion. For ancient peoples, motion would usually have constituted riding a horse, running, or walking. The feeling of wind against one's face would immediately have revealed one's motion. Even when being transported in a carriage or cart, the view of the landscape moving by would have been sufficient evidence of motion. However, no such evidence for the motion of the Earth was felt or seen.

    Using the estimated distance between the Earth and the Sun, determined using calculations of Aristarchus and Eratosthenes, one finds that if the Earth was orbiting the Sun, it would have been moving through space at a brisk 5200 km/hr, or nearly 1.5 km/s (we now know the actual speed to be much faster!). It stood to reason that such dramatic speed would be immediately noticeable, and that it should even cause a person to be swept off their feet. With no such experience or observation, it was concluded that the Earth could not possibly be moving.

2. **If the Earth was moving in a circular orbit, the stars should exhibit back-and-forth motion as we view them from slightly different angles.** Hold your left arm out all the way, with your thumb up. Close one eye and hold your right arm halfway out with your thumb up. With that one eye closed, line up your two thumbs. Once you have them lined up, switch eyes without moving your thumbs. By doing this, you will notice that the thumb closer to your face appears to have shifted with respect to the more distant thumb. This effect is known as *parallax*.

When your viewing position changes, you see nearby objects projected in different positions against a more distant background. In the case of the Earth, the Greeks recognized that an Earth in motion should see certain nearby objects shift, as depicted in **Figure 2.17**, against the background objects. However, no such shift was observed in any celestial objects.

3. **Since things ought to tend toward the center of the Universe, if the Sun was at the center, then we would all fly off the Earth.** Having no functional understanding of the concept of gravity, the Greeks believed only in a force that tended to draw things toward what they believed to be the center of the Universe. It was not until Galileo Galilei, some 1900 years later, that the notions of gravity were ever conceived or studied.

Aristarchus argued that no parallax was observed because the stars are exceedingly distant, thus reducing the angle in **Figure 2.17** to one

> **Parallax**
> The apparent shifting of nearby objects with respect to a distant background when viewed from different locations.

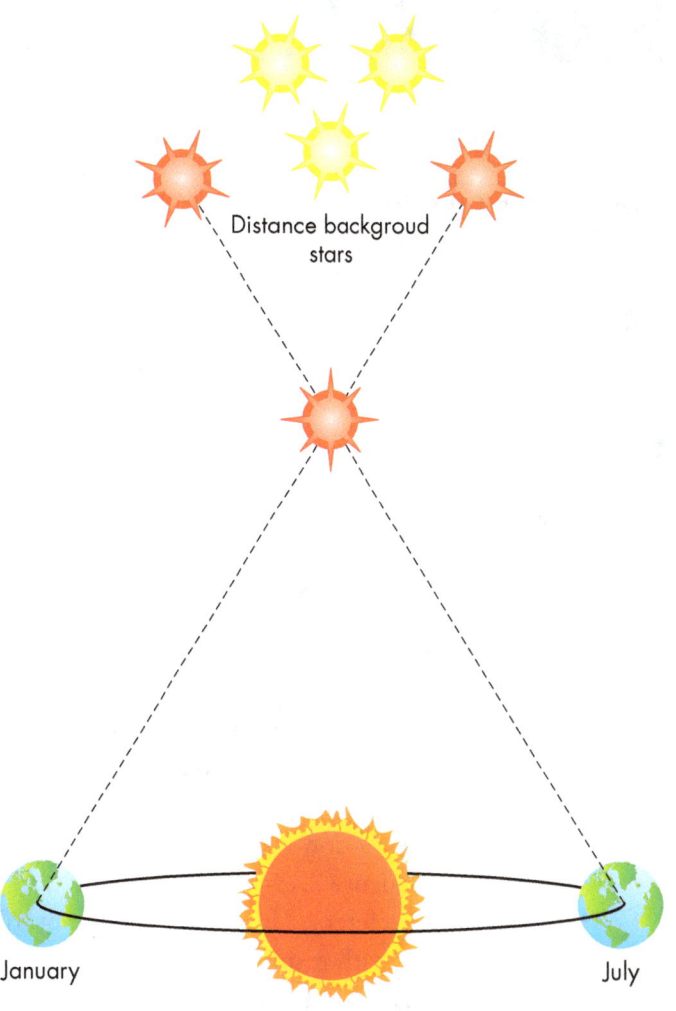

Figure 2.17. As the Earth orbits the Sun in the heliocentric model, we ought to observe relatively nearby stars (red) shift from one position against the background, viewed in January, to another position six months later (July). The apparent shift of a nearby object with respect to more distant background objects is called parallax.

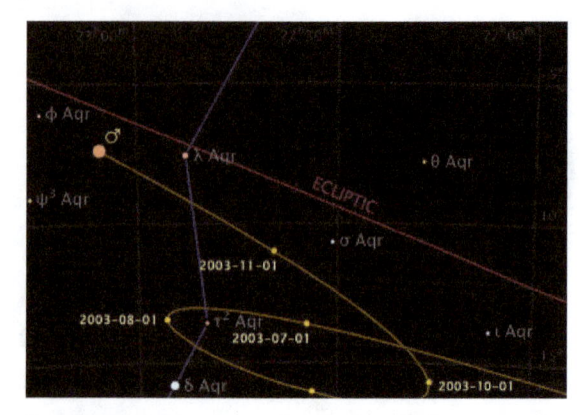

Figure 2.18. An example of retrograde motion displayed by Mars against the background stars during 2003.

Copyright © 2008 by Eugene Alvin Villar / Wikimedia Commons / CC BY-SA 3.0

> **Retrograde motion**
>
> The temporary reversal of a planet's normal west-to-east motion against the background stars.

that is immeasurably small. However, this was dismissed because it produced a Universe of unfathomable extent.

While most observations and experiences seemed to support the notion of a geocentric Universe, there was one detail in the motions of the planets that initially seemed to defy explanation. The motions of some planets against the background stars proceeded from west to east, while others moved eastward away from the Sun for a while, then turned around and moved back toward the Sun again. They would then continue moving westward past the Sun for a bit, and then reverse course again. This behavior, exhibited by Venus and Mercury, differed from Mars, Jupiter, and Saturn in that they always remained near the Sun in the sky, while the latter three were not "bound" in the same way. However, Mars, Jupiter, and Saturn also demonstrated brief periods of backwards motion, shown in **Figure 2.18**. This unpredictable motion was bizarre and could not easily be accounted for in a simple geocentric model with circular orbits around the Earth.

This temporary reversal of the otherwise normal west-to-east motion is called *retrograde motion*. Why should the planets deviate from their supposedly perfect circular orbits? The greatest minds of the time set about trying to solve this problem and it was ultimately a second-century AD astronomer named Ptolemy (90–168 AD; shown in **Figure 2.20**) who published a model that attempted to explain it.

In Ptolemy's model, he started with the Earth at the center. Since the Moon and Sun moved in predictable, smooth motions, he used perfect circles to indicate their orbits around the Earth. Each planet was affixed to a circle he called an epicycle. The center of this epicycle orbited the Earth by being attached to a larger circle called a deferent. This helped to explain retrograde motion because when the planet moved along the outer part of the epicycle, as shown in **Figure 2.19**, it was moving in the same direction as the overall motion of its epicycle along the deferent. However, when it moved along the near part of the epicycle, it was moving backwards with respect to the motion of the epicycle along the deferent, giving it an apparent backwards motion against the more distant stars.

It became clear that while this explained planetary retrograde motions, it did not accurately predict planetary positions in the sky. To help fix this, Ptolemy shifted the Earth away from the center of the deferent, which became known as the eccentric, and balanced it out with a point called the equant. The influence of the equant contributed to the planet's apparent change in speed as it completed an orbit around the Earth. Furthermore, when the planet on its epicycle

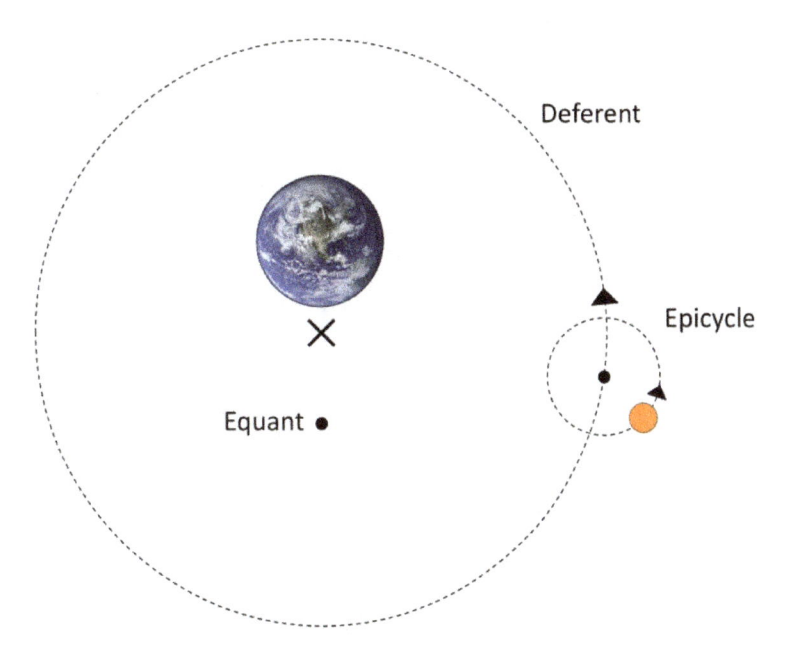

Figure 2.19. The Ptolemaic model of the Universe placed each planet on a circular path called an epicycle, which in turn moved along a circular path around the Earth called a deferent. The Earth was shifted away from center, balanced out by a point called the equant, as a way of explaining the differing angular speeds of a planet as it completed an orbit.

was at its nearest point to Earth, it would appear brighter due to its closer proximity—which was exactly what was observed.

Mercury and Venus never stray far from the Sun in the sky, with Mercury's maximum elongation (its largest angular separation from the Sun) being less than 30 degrees, and Venus' maximum elongation being about 47 degrees. For this reason, a line was drawn to connect the Earth and the Sun, along which were attached the epicycles of Mercury and Venus. Objects were ordered in distance from Earth depending on their apparent speed through the sky: Moon, Mercury, Venus, Sun, Mars, Jupiter, and Saturn. The stars were fixed on an outer surface that rotated around Earth as well.

This system of circles upon circles could then be tuned to reproduce the apparent motions of the planets. Once completed, the Ptolemaic model provided the most accurate way of predicting the positions of the planets that had yet been developed—even more accurate than Aristarchus' heliocentric model! However, it was still limited to an accuracy of only about 5 degrees, which corresponds to roughly half the width of your fist held at arm's length.

Figure 2.20. A medieval painting depicting the astronomer Claudius Ptolemaeus.

Wikimedia Commons / Public Domain

It is worth noting that it is still debated whether or not the Greeks believed this to be an accurate *physical* depiction of the Universe—they may or may not have believed there to be physical deferents and epicycles existing in space. They would surely have recognized the existing inaccuracies of the model and therefore probably would have assumed there was something more they were missing. The model's greatest success was that it qualitatively agreed with what was observed—to a fair degree of accuracy—and also met their expectations about what they thought the primary characteristics of the Universe ought to be (e.g., circular orbits, Earth at the center, etc.). For these reasons, the Ptolemaic model was very highly regarded and was adopted as the standard model of the Universe for nearly 1,500 years.

CHAPTER THREE

An engraving by an unknown artist that first appeared in *L'atmosphère: météorologie populaire* (1888), by Camille Flammarion, depicting a Renaissance-era explorer. The image is commonly interpreted to illustrate the spirit of science—peering beyond that which is known to discover the unknown.

Camille Flammarion / Public Domain

Renaissance Astronomy

Learning Objectives

In this chapter you will learn:
- » How Renaissance thinking/science changed our view of our place in the Universe
- » Who the major players in this intellectual revolution were, and what each contributed
- » What factors influenced the time it took to transition from geocentric to heliocentric thinking
- » How the invention of the telescope fundamentally impacted the field of astronomy
- » How Galileo's observations were interpreted to support a heliocentric model

Key Words
- Astronomical Unit
- Orbital period
- Semimajor Axis

3.1 The Copernican "revolution"

Following the collapse of the Western Roman Empire in the 5th century A.D., much of Europe fell into a lengthy period where little intellectual or cultural advancement occurred. This time period, which by some estimates lasted nearly one thousand years, came to be known as the Middle Ages, or Dark Ages. While the rising influence of intellectual Christianity allowed much of the early Greek scholarship to be preserved, it was not developed in the West any further during this time.

Meanwhile, Islamic influence in the East gradually grew. Territories once held by the Roman Empire (**Figure 3.1**) were conquered and assumed into the growing Islamic Empire (**Figure 3.2**). Through the process of rediscovering the accomplishments of the earlier Greeks, Islamic

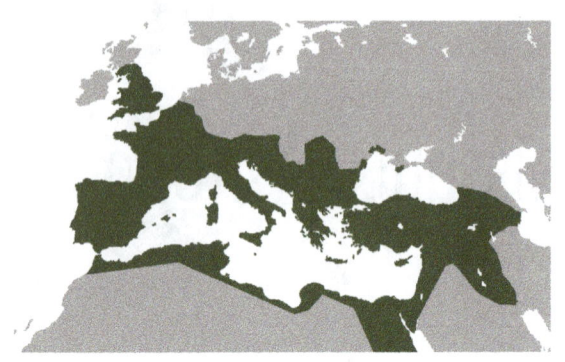

Figure 3.1. The Roman Empire at its greatest extent, in 117 AD

Copyright © 2007 by User:Geuiwogbil / Wikimedia Commons / CC BY-SA 3.0

Figure 3.2. The Islamic Empire by 750 AD

Copyright © 2009 by User:Gabagool / Wikimedia Commons / CC BY 3.0

philosophers nurtured and cultivated their own culture of scientific inquiry. Massive libraries were built and Greek texts were translated into Arabic. As the empire expanded, Arab scholars developed an enhanced understanding of geometry and spherical trigonometry. One of the most valuable contributions that these scholars made to math and science was the introduction of the Arabic numeral system, from which we get the sequence of digits 0–9 used today along with the system of representing a larger number as a series of these digits read left to right.

The Arabs helped to preserve that which had been learned by the Greeks while making considerable contributions to instrumentation and observation as well. As Europe transitioned out of its period of intellectual stasis, the knowledge maintained by the Islamic Empire was then exported back to the Western world in the 11th century. Arabic libraries were translated into Latin, and the Western world took a renewed look at history's greatest intellectual accomplishments.

Two factors came into play that had a dramatic impact on how this "new" knowledge was received. First, the Christian church in Europe was a major player in determining truth

Figure 3.3. A 16th century Ptolemaic depiction of the celestial orbs, centered on the Earth

Peter Apian / Public Domain

from fallacy. Discoveries and predictions needed to mesh with biblical interpretations or else they were dismissed as error or, in the most severe case, heresy. Second, European scholars were hesitant to question the writings of the revered ancient Greeks. Much of what was recovered during this time was widely accepted as fact, with the assumption that it had all been well studied previously. This led to the continued acceptance of Ptolemy's geocentric model throughout the medieval period (**Figure 3.3**).

With the blossoming of European scholarship came the development of universities and the printing press. Suddenly not only were scholars able to work together in a newly focused and collaborative way, but they were also able to reproduce and distribute their thoughts with increasing ease. This made possible the rapid spreading of new ideas and debate, ultimately allowing the process of discovery to proceed at an increasing rate.

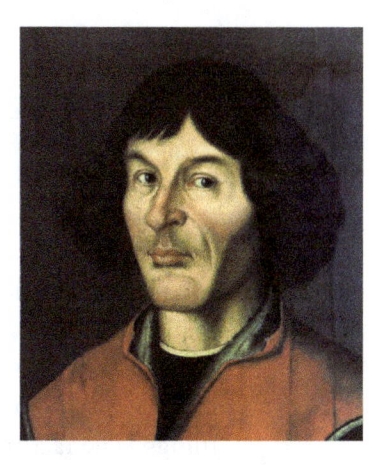

Figure 3.4. Nicolaus Copernicus
Wikimedia Commons / Public Domain

As scholarship grew, criticisms of the ancient Greek works arose—albeit rarely. Individuals studying the geocentric model came to realize that not only was it not as accurate as they would have liked, but it seemed frustratingly complex. Even Ptolemy himself had been credited with stating, "We consider it a good principle to explain the phenomena by the simplest hypothesis possible." This principle, echoed in sentiment by philosophers and scientists throughout the centuries that followed (and which would later come to be known as Occam's Razor), became increasingly fundamental in the evaluation of models and hypotheses.

Such was the motivation behind the work of Nicolaus Copernicus (1473–1543 AD; shown in **Figure 3.4**), a Polish lawyer, mathematician, and astronomer. Troubled by the complexity and seeming incoherence of the Ptolemaic model, in particular its use of the equant, for which there was no physical analog, Copernicus desired something more elegant and more physically plausible. Following his studies first at Cracow University and later in Italy, he went to work developing a more adequate theory.

As Copernicus studied the Ptolemaic model, and the writings of those who had studied it previously, he was struck by the fact that the calculation of each planet's position in the sky relied upon the Sun's motion as well. With this insight, Copernicus wrote a manuscript entitled *Commentariolus* (or *Little Commentary*), which he circulated to select colleagues between 1512 and 1514. While omitting the mathematical analysis, Copernicus used this manuscript as a means of presenting several fundamental axioms of a heliocentric (sun-centered) model of the Universe:

1. Celestial objects do not all orbit a single point.
2. The center of the Earth is not the center of the Universe, but only the center of the Moon's orbit around the Earth.
3. The Sun is near the center of the Universe, and all the planets orbit around it.

4. The distance from Earth to the Sun is insignificant compared to the distance to the stars.
5. The stars are fixed, and their apparent daily motion (and that of the Sun) is due to the rotation of the Earth.
6. The apparent motion of the Sun against the background stars throughout the year is due to the Earth's orbit around it.
7. The apparent retrograde motion of the planets in the sky is due to the Earth's orbital motion around the Sun.

These statements formed the backbone of the Copernican model. To many of those who read it at the time (of whom there were few), these helped to immediately explain several things. The fact that stellar parallax had not been observed was not because the Earth was not moving, but because the Universe must be a much larger place than had originally been conceived.

In **Figure 3.5**, you can see that if you took the foreground star (red) and moved it extremely far away from the Sun, as its distance increases the parallax angle (p) shrinks in size. When the star is extremely far away, the angle is immeasurably small. Despite the Earth's orbital motion changing the observer's position, the parallax angle is just too small to observe.

The assertion that the Earth and all other planets are moving also helped provide an elegant explanation for the observed retrograde motion of planets. The Ptolemaic model had leaned on the use of epicycles to explain why a planet might sometimes appear to briefly reverse direction against the background stars. **Figure 3.6a** shows that this reversed motion occurs when the planet is moving along the near side of its epicycle.

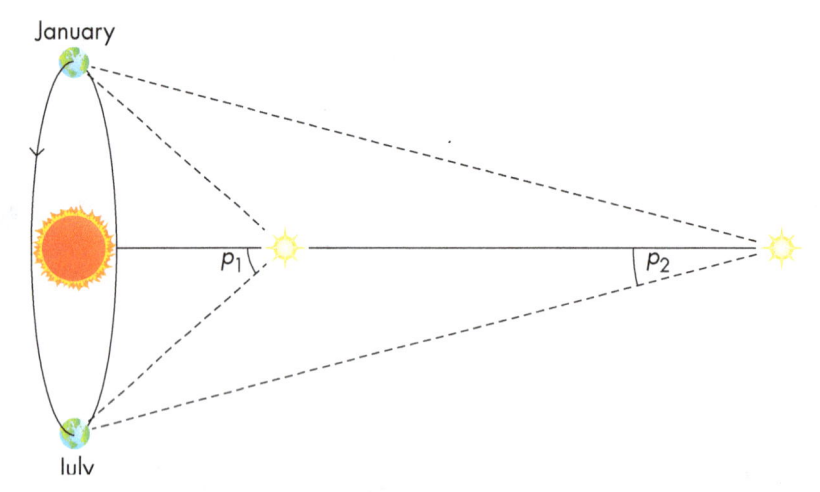

Figure 3.5. When considering stars at increasing large distances, the parallax angle p changes from large (p_1) to smaller (p_2). If the distance is increased enough, p becomes too small to observe.

In the Copernican model, because our planet is moving along an orbit whose path is shorter than that of the more distant planets like Mars and Jupiter, we periodically pass these planets as we revolve around the Sun. **Figure 3.6b** shows how this passing motion produces the observation that the outer planet seems briefly to be moving backwards against the background stars.

The same experience occurs when you drive on the highway. Suppose you are traveling down the road in your car along a highway that has two lanes going in each direction separated by a median. As you catch up to a slow-moving car, you move over to the left lane to pass it. While you are passing the car, from your viewing perspective, that car seems to be briefly moving "backwards" with respect to you and the distant treeline. Only once you have passed the car does it once again regain the appearance of moving forward with respect to the distant treeline.

The manuscript that Copernicus circulated did not draw a dramatic amount of attention. This was likely in part due to the fact that he was a relatively unknown person from a distant corner of Europe. Moreover, there was still no observable proof that the Earth was moving. Recall that the ancient Greeks had originally rejected the heliocentric model because 1) no stellar parallax was observable that might indicate Earth's motion, 2) no wind was felt that might indicate Earth's motion, and 3) there was no alternative explanation for why we might remain stuck to the Earth instead of flying off into space as the Earth moved along. From an observation standpoint, nothing had changed.

Copernicus knew that his work, while outlined in concept, was not complete until he had worked it out in mathematical detail. He withdrew and spent the next thirty years of his life expanding his ideas into a massive book entitled *De Revolutionibus Orbium Coelestium (On the Revolutions of the Celestial Spheres)*. In the culmination of his efforts, he not only reemphasized the natural elegance and physical simplicity of his model, shown in **Figure 3.7**, but also devoted roughly 95% of its contents to the mathematical proof that the model could be used

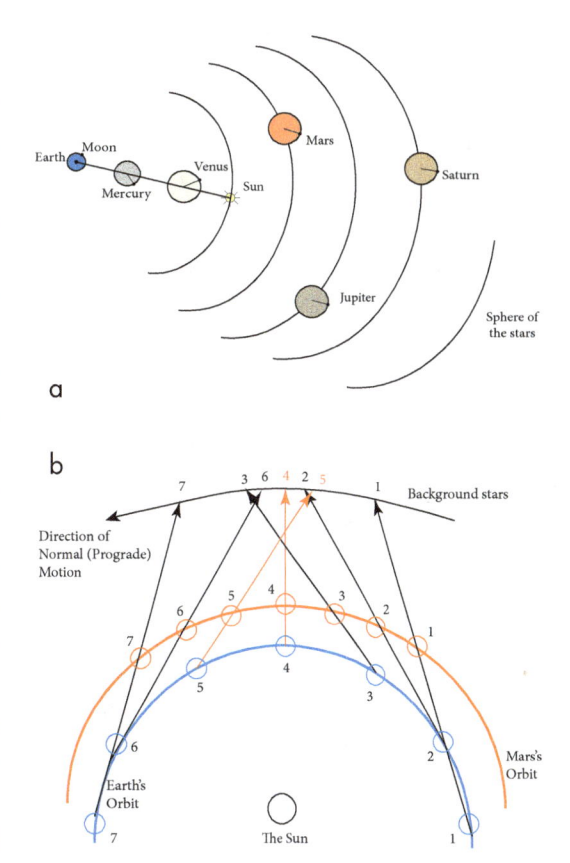

Figure 3.6. In the Ptolemaic model (a), planets follow epicycles as they orbit the Earth. This produces periods when the planet is moving along its epicycle in the opposite direction it had been formerly moving—giving the appearance of retrograde motion. In the Copernican model (b), Earth passes by the outer planets in its orbit and in doing so, has the perspective of watching the outer planet briefly move in reverse against the background stars.

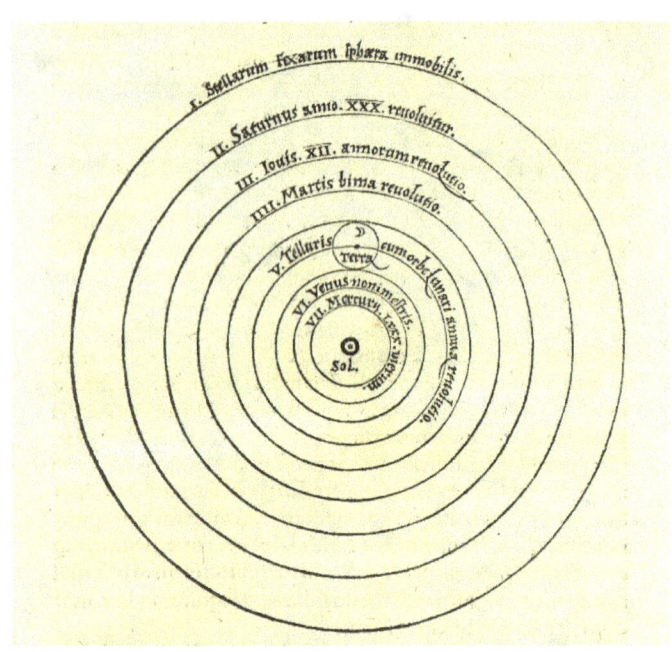

Figure 3.7. The heliocentric model, as drawn out by Copernicus in *De Revolutionibus*. All planets are depicted following perfectly circular orbits centered on the Sun.

Nicolaus Copernicus / Public Domain

to accurately describe the Universe. Having completed his exposition and arranged for its publication, Copernicus could only sit back and await its reception by the scientific world.

Unfortunately, this much grander attempt at getting his ideas across also failed to produce much of a stir. There are several possible reasons for this. First, the mathematical treatment in his work was extremely technical. It would have required a knowledgeable and focused reader to appreciate the author's efforts and while some who read it were intrigued by his ability to eliminate the need for an equant, the mathematical description of the circular, Sun-centered orbit still required "epicyclets" to provide reasonable predictions of planetary positions.

Second, after Copernicus had finished writing it, his publishing assistant had added a preface to the book that effectively dismissed the book's contents. While the motivation for such an addition is likely lost to history, the preface explained that this was just one way of simplifying the calculations and that it made no claim to be physically accurate.

One final reason that it made no immediately significant impact in scientific community was that when one used the heliocentric model to calculate the observed positions of planets in the sky, the results were no more accurate than the geocentric model. The new model did not pass a fundamental test in the eyes of its critics. While extensive research by Dr. Owen Gingerich of Harvard University indicates that many scientists of the day did in fact read De revolutionibus, it still failed to take the world by storm in the way Copernicus hoped it would. Copernicus died soon after it was published and the immediate impact of his work was minimal.

3.2 Refinement of the model

Shortly after Copernicus' death, the next great astronomer was born. Tycho Brahe (1546–1601 AD; **Figure 3.8**) was

Figure 3.8. Tycho Brahe

Museum of National History at Frederiksborg Castle / Public Domain

born into Danish nobility and would use his privileged life to become the most astute observational astronomer the world had yet seen.

Although he had been encouraged by his family to study law at the University of Copenhagen, Tycho became interested in astronomy after observing a solar eclipse in 1560. His dedication grew when he observed a planetary conjunction of Jupiter and Saturn in 1563, when the two planets slowly passed one another in the sky. Recognizing that both the Ptolemaic and Copernican models predicted this conjunction differently, and both incorrectly, Tycho saw an opportunity and began to systematically observe the sky night after night, quickly becoming intimately familiar with the positions of the stars. When the most accurate instruments of the day were insufficient, he built larger ones. Through this process, he began to compile a massive catalog of stellar and planetary positions.

Figure 3.9. Uraniborg, where Tycho Brahe lived and worked for much of his professional life.

Blaeu / Blaeu's Atlas Major / Public Domain

In 1572, Tycho observed the appearance of a new star in the constellation of Cassiopeia, which he referred to as a "nova" (though which we now know was a supernova—an exploding star). His subsequent reports on this and the Great Comet he painstakingly observed in 1577 allowed Tycho to gain a strong reputation as an observer.

It was around this time that King Frederick II of Denmark financed an observatory for Tycho on the island of Hven, off the Danish coast, and gave him control over the island. The observatory Tycho built, which he named Uraniborg (Heavenly Castle), shown in **Figure 3.9**, not only became the central location for Tycho's work but was also the center for his rather boisterous social life. Its observation turrets provided Tycho with unimpeded views of the heavens using cutting-edge instruments, and its expansive dining halls and rooms provided the setting for a multitude of raucous parties. The accounts of these parties include a story-telling dwarf named Jepp, whom Tycho retained to amuse his guests, and a pet elk that was said to roam freely throughout the castle and that Tycho would rent out to friends.

Based on his keen observations of the motions of planets in the sky, Tycho set about creating his own model of the Universe. He did so because while he acknowledged that the Ptolemaic model produced dramatic errors in predicted positions, he was opposed to the Copernican system as well. This was because despite his careful measurements, no stellar parallax could be detected and this implied that the Earth was not moving. However, he did recognize that Mercury and Venus must orbit the Sun. The Tychonic model, as it came to be known, sought to combine popular elements of both the Ptolemaic and Copernican models.

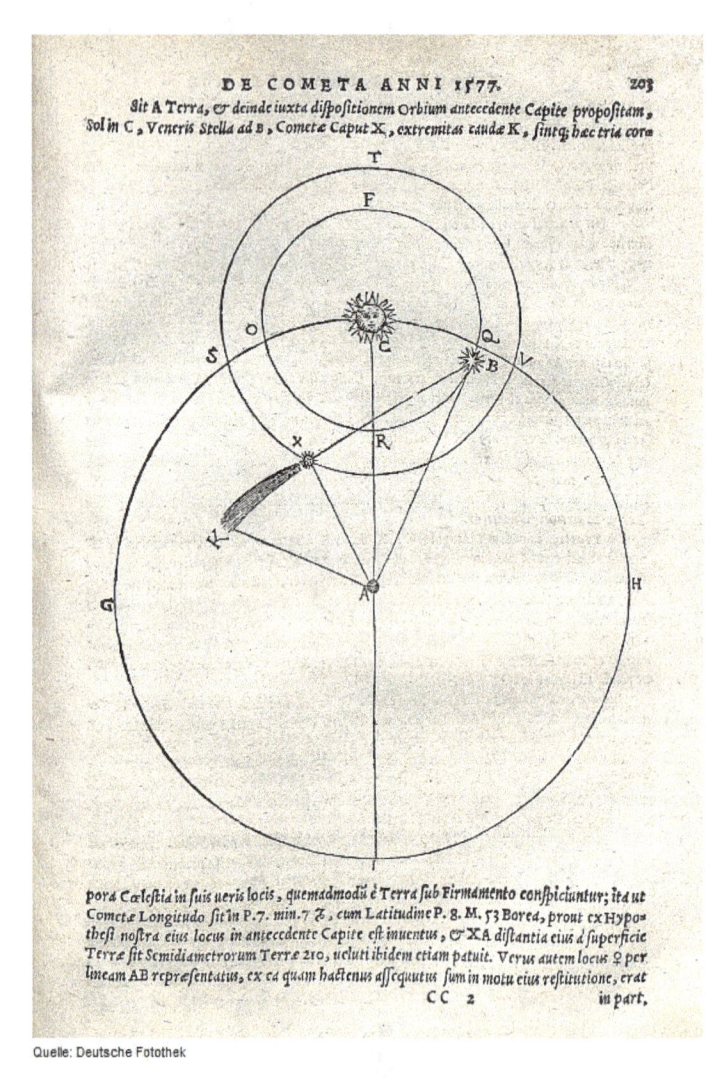

Figure 3.10. The Tychonic model of the Universe, where every planet but Earth orbits the Sun, and the Sun orbits the Earth.

Tycho Brahe / Deutsche Fotothek / Public Domain

Figure 3.10 depicts a diagram of Tycho's model, laid out in his book *De mundi aetherei recentioribus phaenomenis* (*Concerning the New Phenomena in the Ethereal World*). In his model, every planet but Earth orbited the Sun. This was done to avoid the inclusion of non-physical equants and to elegantly explain retrograde motion. The Earth itself was fixed in the center of the model and the Sun, with its family of planets, revolved around it. Although this model was interesting and attempted to be a compromise between the two competing models of the day, it did not initially provoke a tremendous amount of discussion in the scientific world, primarily because Tycho lacked the mathematical skills to expound upon it.

When Christian IV ascended to the Danish throne, Tycho fell out of favor with the royal court. With his funding revoked, Tycho packed up his instruments and his family and moved to Prague, where the Emperor Rudolph II gave him a high appointment and a new observatory. Tycho continued his observations there, adding to his catalogs of planetary positions.

Distressed as he was that his observations might go unused upon his death, this move to Prague was extremely fortuitous. It was there that Tycho took on a young assistant from Germany named Johannes Kepler (1571–1630 AD), who had a strong mathematical background and a keen interest in astronomy, being an adherent to the Copernican model. Kepler's interest in astronomy, inspired by witnessing the Great Comet in 1577, was hindered by the fact that his vision was terrible. Together, they were indomitable.

This partnership was short-lived, however. Less than a year into their working relationship, Tycho became ill following a long night of drinking at a party. Evidently his strict adherence to etiquette forbade him from leaving the table, despite feeling "the tension in his bladder increase" (as Kepler recorded). Modern evaluations say he died from either a burst bladder or uremia.

Figure 3.11. Johannes Kepler, later in his life.

Wikimedia Commons / Public Domain

Following Tycho's death, Kepler (shown in **Figure 3.11**) now had free access to the massive quantities of data that Tycho had been closely guarding. Applying his mathematical expertise to Tycho's observational data for Mars, Kepler was certain he could fix the inaccuracies in the Copernican model. This endeavor would take him eight years.

In his quest to solve the problem, he insisted that not only should the mathematical model explain the observations qualitatively (that is, in their general appearance) but they should also agree quantitatively. In his mind, there was no good reason that the model should produce predicted positions that were so far from the real values. For the model to ultimately work, he had to make a major leap away from an assumption inherent to both the Ptolemaic and Copernican models—the circular shape of planetary orbits.

It had long been assumed that planets followed circular paths through space, dating back to the ancient Greeks and their love affair with the perfect circle and uniform circular motion. Copernicus had set up his model with perfectly circular orbits, nearly concentric on the Sun, with movement at constant velocity. However, Kepler realized that a circular shape did not fit the data points measured by Tycho. In order to produce a path that *did* fit the data within the strict limits on accuracy he imposed, he had to adjust the shape to that of an ellipse.

An ellipse is a curved shape that can, in extreme cases, resemble an oval, shown in **Figure 3.12**. The degree to which the ellipse is stretched out is called its *eccentricity*. A circle is an ellipse with zero eccentricity, and the more stretched out the ellipse gets, the higher its eccentricity—up to a maximum of 1. Within an ellipse there exist two points called *foci* (plural of "focus"). The separation of the two foci is determined by the eccentricity, and in the case of a circle, the two points coincide. In any ellipse of non-zero eccentricity, the foci are offset from the center of the ellipse by increasing amounts as the eccentricity increases.

In the case of objects in our solar system, Kepler noticed that not only was Copernicus' assumption regarding perfectly circular orbits incorrect, but his assumption about the Sun being at the center was also wrong. In the case where objects in our solar system follow elliptical orbits, Kepler found that the Sun is located at one focus (there is nothing at the other focus). Most of the planets in our solar system follow orbits that are very close to being circular (example *c* in **Figure 3.12**), which was why Copernicus' assumption had still produced results that were fairly close to correct. Comets, on the other hand, follow very eccentric orbits (example *a* in **Figure 3.12**) that take them very near to the Sun at their closest approach, called perihelion, and very far away from the Sun at their most distant point, aphelion.

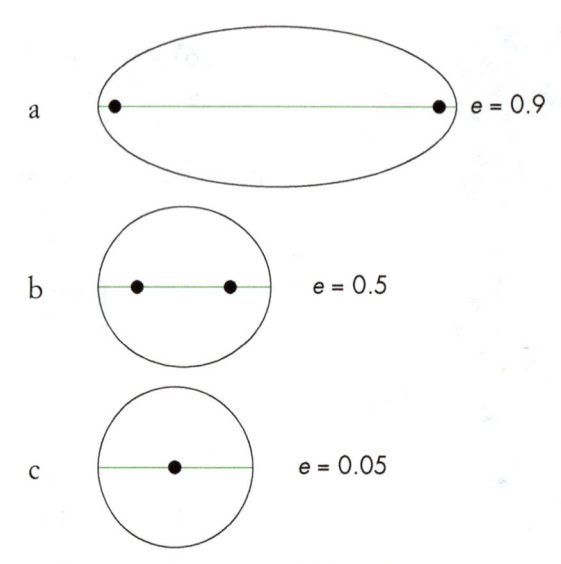

Figure 3.12. Three examples of ellipses, where ellipse (a) has the greatest eccentricity and greatest separation distance between foci, while ellipse (c) has zero eccentricity and the two foci coincide. The green line drawn along the length of the ellipse is called the major axis.

Once Kepler dismissed these two key ideas from Copernicus, he was able to finally solve the problem and produce a model that correctly predicted the observed positions of planets in the sky. In doing so, he developed over the span of a decade what have come to be known as Kepler's Laws of Planetary Motion:

1. Planets follow elliptical orbits around the Sun, and the Sun is at one focus.
2. Planets move more quickly in their orbit when they are closer to the Sun and more slowly in their orbit when they are farther from the Sun.
3. The time it takes a planet to complete one orbit around the Sun, called the *orbital period (p)*, is directly related to the size of the planet's orbit, indicated by the *semimajor axis (a)*, by the equation $p^2 = a^3$, when the period is measured in years and the semimajor axis is measured in astronomical units.

Orbital period
The time required for an object to complete one orbit around another object.

Semimajor Axis
Half of the length of the longest diameter of an ellipse. This also represents the average distance of a planet from the Sun.

While Kepler had no understanding about *why* these laws of planetary motion should be true, he could see mathematically that a planet's position changed much more dramatically when it was closer to the Sun than when it was farther away, indicating that it was speeding up on approach and slowing down as it moved away (**Figure 3.13**). He presumed that some compulsion must exist for the Sun to pull on a planet, with strength that dropped with increasing separation distance, but did not investigate this idea.

He noticed that there was an elegant relationship between the orbital period p (the time it takes a planet to complete one orbit around the Sun) and the semimajor axis a (half the major axis shown in **Figure 3.12**) for every planet. In those days, absolute distances throughout the solar system were not known in physical units; all that was known were relative distances based on the standard of the time—the Earth's distance from the Sun. This distance from the Earth to the Sun was called the *astronomical unit* (AU) and was, by definition, 1 for the Earth. Relative distances from Earth to the other planets in AUs were also known. As Kepler interpreted Tycho's data from a heliocentric perspective, he saw that the semimajor axis of any planet, which is effectively that planet's average distance from the Sun over one complete orbit, when cubed, was equal to the orbital period squared. Being that Kepler was working without physical distance units, this relationship was only true when the distance was in AUs and the period in years.

Applying his laws of planetary motion, Kepler was able to accurately predict the positions of Mars in the sky. His accomplishment was a huge success, and a leap forward for the field of

> **Astronomical Unit (AU)**
> The average distance between the Earth and the Sun. The Earth is 1 AU from the Sun.

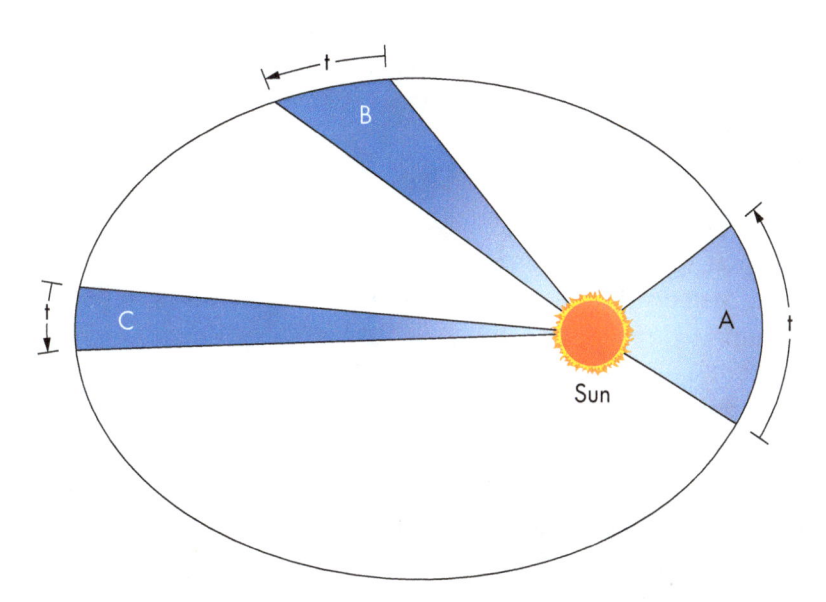

Figure 3.13. This diagram illustrates Kepler's 2nd law of planetary motion. As he described it, a line connecting the Sun and a planet sweeps out equal areas in equal time intervals. Physically, this means that over some time t (for example, one month) a planet might move along the arc of its orbit tracing out area A, while over the same time interval t the planet would move along a much shorter arc tracing out area C. This indicates that if both time intervals are the same, but the planet moves a greater distance along its orbit in A than C, it must be moving faster at A than at C.

Copyright © 2011 by User:RJHall / Wikimedia Commons / CC BY-SA 2.0 AT

> ## Tools of the Trade 3.1
>
> While Kepler's Third Law may look intimidating at first
>
> $$p^2 = a^3$$
>
> we can easily use it if we have a basic scientific calculator. Suppose we know that the orbital period of Mercury is 88 days—that is, it takes 88 days for Mercury to go once around the Sun. We first convert that into years:
>
> $$p = \frac{88d}{365d/yr} = 0.241 yr$$
>
> Recall that if we insert the period (in years) into Kepler's Third Law, we will end up getting the semimajor axis, the average distance from the Sun, in AUs. Squaring the period we get
>
> $$p^2 = (0.241)^2 = 0.0581$$
>
> This means that
>
> $$a^3 = 0.0581$$
>
> so we can then solve for a:
>
> $$a = (0.0581)^{1/3} = 0.387 AU$$
>
> Therefore, Mercury's average orbital distance from the Sun is 0.387 astronomical units.

science. Combining his mathematical skills with observational data and theoretical modeling, Kepler provided an excellent example of the modern scientific method.

Kepler devoted eight years to developing the first two laws of planetary motions and applying them to the orbit of Mars. He published them in a massive book entitled *Astronomia nova (A New Astronomy)* and spent ten subsequent years developing his third law of planetary motion. However, the reception to his publication was cool at best. While philosophers and religious leaders acknowledged its accuracy in making calculations, they still rejected the notion of a heliocentric Universe. Others also rejected the notion of orbits that were not perfectly circular and entreated him to retain circular orbits with small epicycles for correction. Kepler was shocked! With dismay, he turned his attention away from that on which he had spent so much time and effort to accomplish and focused on other subjects for the remainder of his career. It would take an observer to finally begin to chip away at the notion of a geocentric Universe.

Figure 3.14. Galileo Galilei, considered by many to be the father of experimental physics.

Justus Sustermans / Public Domain

3.3 Galileo and his telescope

Galileo Galilei (1564–1642 AD), shown in **Figure 3.14**, was an Italian scientist and mathematician who had early on come to accept the Copernican model of heliocentricity, albeit with reservation at first. While visiting Venice in 1609, he heard about a device called a telescope, constructed originally in Holland by a lensmaker named Hans Lippershey, which used two glass lenses mounted in a tube to magnify distant objects and make them easier to observe. Fascinated by the prospect, he returned home and learned how to grind and polish glass into lenses—thus creating his own telescope. He quickly became skilled at it and within the year he had fashioned for himself a telescope with a magnifying power of 30X, making distant objects appear thirty times larger than they would appear with the naked eye alone. He then turned his instrument to the night sky.

While Galileo did not invent the telescope, nor was he even the first to view the night sky, he was the first to perform such observations in a systematic way and subsequently record and publish his results. His first set of observations appeared in a small booklet entitled *Sidereus Nuncius (Sidereal Message)*. This publication instantly caught fire and rapidly spread throughout Italy and Europe.

One of Galileo's first reports was on his observations and engravings of the lunar surface. Long believed to be perfectly spherical, telescopic magnification revealed it to be a fantastic world of craters, mountains, valleys, and plains, shown in **Figure 3.15**. As the terminator, the line separating daylight from darkness on the Moon, moved along it revealed a myriad of features previously unseen. All of Europe was captivated by his drawings.

Additionally, Galileo's telescope revealed that the hazy Milky Way was actually composed of countless stars too faint to be seen or resolved with the unaided eye. This suggested that the full extent of the Universe, where stars were thought to be fixed to an outer sphere, may actually be much greater than previously thought.

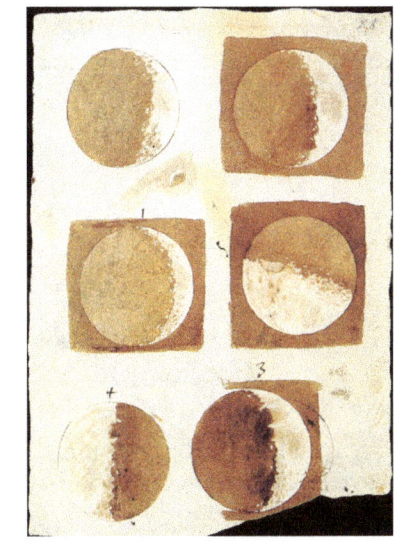

Figure 3.15. Phases of the Moon sketched by Galileo and published in his *Sidereus Nuncius*. His sketches revealed a huge variation in topographical features, thus refuting the idea that the Moon was a perfect sphere.

Galileo Galilei / Public Domain

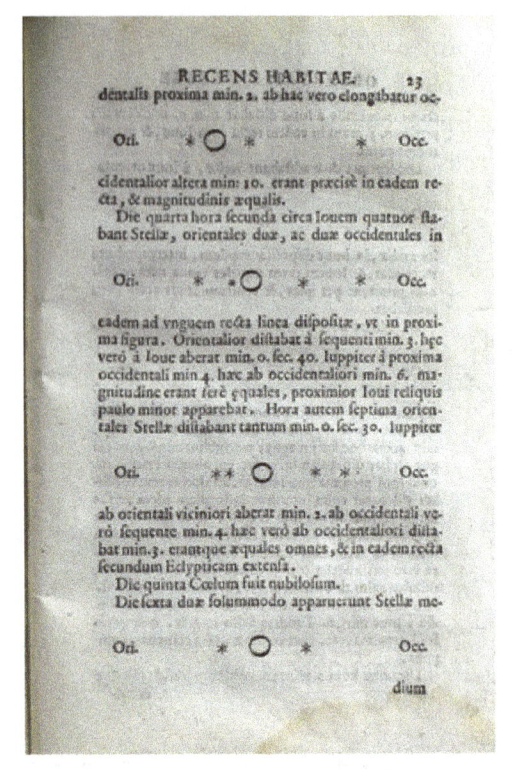

Figure 3.16. Galileo's sketches of the motions of the mysterious stars around Jupiter.

Perhaps the greatest revelation published in his first work was of the existence of four "stars" that seemed to be orbiting around Jupiter. As he had observed Jupiter from one night to the next, he noticed the presence of four small points of light nearby. With each night's observations, he began to notice that those four points of light never deviated far from Jupiter and followed it across the sky. Furthermore, their motion in the vicinity of Jupiter was periodic. He puzzled over whether he was observing Jupiter move with respect to background stars, or watching stars move with respect to Jupiter, ultimately deciding on the latter. These stars, which he later referred to as "planets," must be orbiting around Jupiter.

This announcement was huge. In the standard geocentric model, every object in the Universe orbited the Earth. With his sequence of observations of Jupiter, taken over more than 60 nights spanning several months, Galileo demonstrated that there did exist objects that did *not* orbit the Earth. In his discussion, he stated,

… our own eyes show us four stars which wander around Jupiter as does the moon around the earth, while all together trace out a grand revolution about the sun in the space of twelve years.

Sidereus Nuncius, 1610

In one fell swoop, he declared that his observations appeared to support the heliocentric model. As all of Europe received his reports and allowed them to sink in, Galileo continued his observations. Turning his instrument to Saturn, he noticed a curiosity: this planet seemed to have company too, though in Saturn's case it was accompanied by two stars that hugged it, one on each side (**Figure 3.17**). Strangely, just a couple years later these stars had completely vanished. Then, oddly enough, when Galileo observed it again a couple years later they had reappeared and looked to him like "ears" on Saturn. While we now know that what Galileo was observing were Saturn's rings, and that when they seemed to vanish he was simply seeing them edge-on, these early

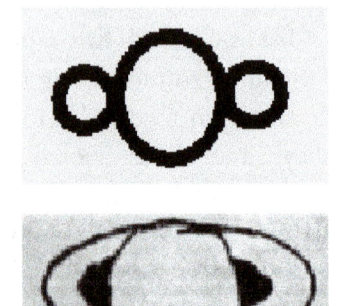

Figure 3.17. Sketches of Saturn made by Galileo in 1610 (top) and 1616 (bottom).

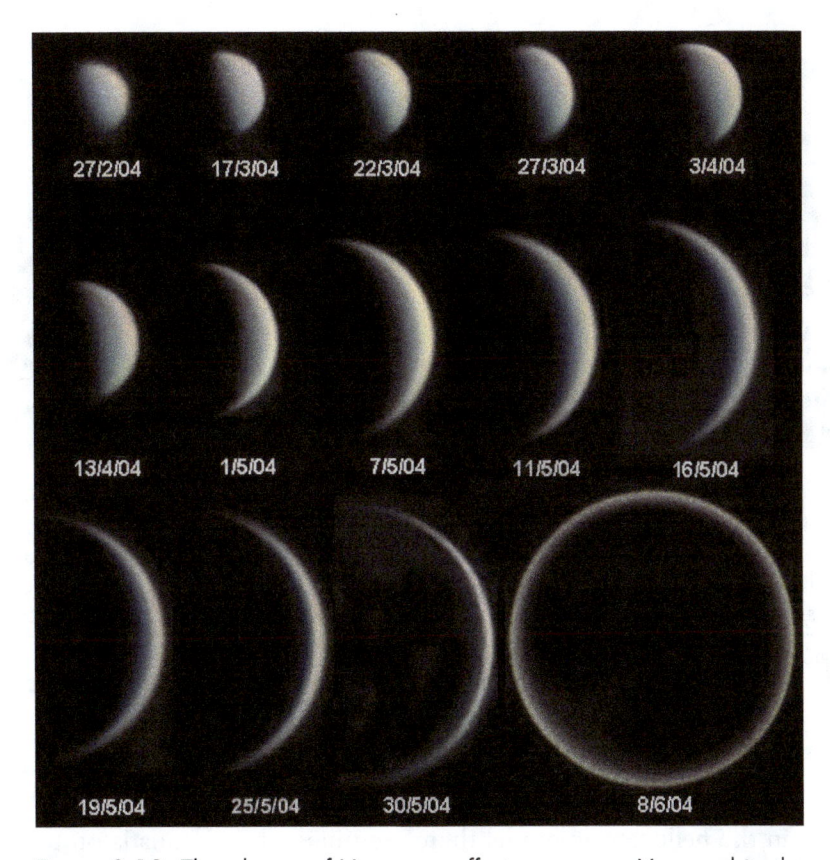

Figure 3.18. The phases of Venus, an effect we see as Venus orbits the Sun. Galileo also witnessed the progression of Venus' phases as the planet moved from behind the Sun (top left) to in between the Earth and Sun (bottom right).

Copyright © 2005 by Statis Kalyvas - VT-2004 Programme.

observations of our ringed planet with instruments of relatively low quality were particularly baffling to astronomers for the next several decades.

Turning his attention to Venus, Galileo made one final series of observations that again caught fire throughout Europe. As Galileo observed the appearance of Venus over several months he noticed that it went through phases, similar to those the Moon displays, as shown in **Figure 3.18**. Phases occur when we observe different portions of the surface of a celestial object illuminated by the Sun. When the Moon is in its new phase, we see no portion of the sunlit side because the Moon is in between Earth and the Sun, meaning the sunlit side is on the far side of the Moon. However, when the Moon is full, it has completed half an orbit around the Earth, meaning that it is now on the opposite side of the Earth as the Sun. The side lit by the Sun now fully faces the Earth.

It was not unexpected that Venus should go through phases. In the Ptolemaic model, Venus orbited a point in space directly between the Earth and the Sun. This kept it close to

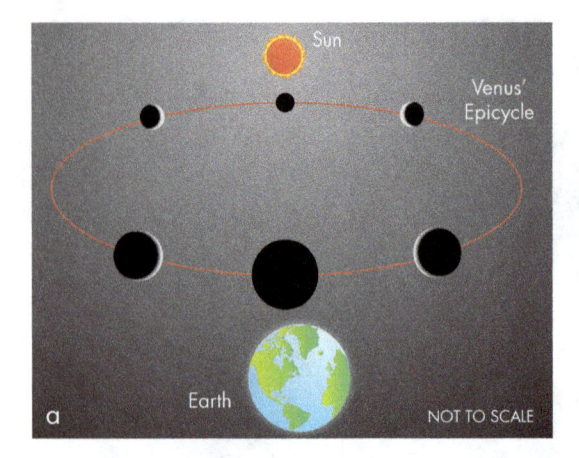

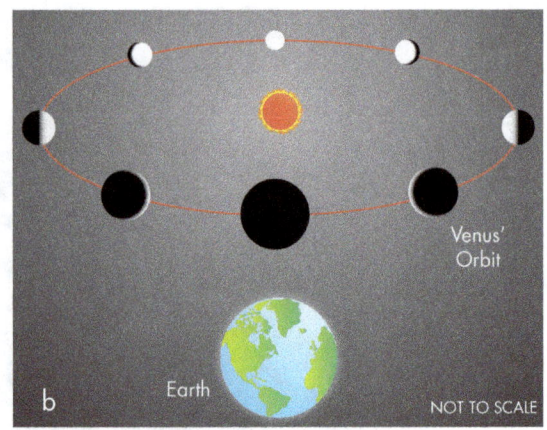

Figure 3.19 (a). The sequence of Venus' phases according to the Ptolemaic model, where Venus was always in between the Sun and the Earth. All that would be seen are a series of crescents, divided by recurring new phases. (b) The sequence of Venus' phases according to the Copernican model, where Venus is orbiting the Sun. The observation of gibbous and full phases indicates that Venus is, at times, on the opposite side of the Sun as the observer on Earth, thereby supporting the heliocentric model.

(left) Astrobryguy / Wikimedia Commons / Public Domain; (right) Sagredo / Wikimedia Commons / Public Domain

the Sun in the sky at all times, and would have produced a sequence of phases that progressed from new, along a series of crescents, then back to new, shown in **Figure 3.19a**.

However, in the heliocentric model there are times when Venus is on the opposite side of the Sun as the Earth (**Figure 3.19b**), meaning that the observer should be able to see the sunlit portion of Venus take on an appearance beyond half-lit—a phase that is referred to as a gibbous. Such a gibbous phase would not appear in the Ptolemaic model, so the observation of such a phase would be solid evidence in support of Copernicus.

This is exactly what Galileo observed and reported in 1613. His report, entitled *Letters on Sunspots* (which also detailed his observations of Saturn and his observations of dark spots on the Sun), was widely read throughout Europe and it wasn't long before the adherents to the Ptolemaic model began to gradually shift. Unfortunately, the shift was not primarily toward the heliocentric model, but toward the Tychonic model. Observations of Venus' phases could support both the Copernican model and the Tychonic model, and since there still existed the belief that the Earth remained stationary, the Tychonic model became favored. Galileo, on the other hand, dismissed Tycho's model as nothing more than rubbish and compromise.

At this point, the Church stepped in. While acknowledging that the Sun-centered model did indeed provide a means for accurate calculation of planetary positions in the sky, and acknowledging the intriguing observations of Galileo, a committee in 1616 formally declared heretical the notion that the Sun, not the Earth, was the center of the Universe. Copernicus' book *De revolutionibus* was banned.

In 1623, an old friend of Galileo's was elected to be the new pope—Urban VIII. Believing he now had the support of the Church, Galileo set about writing a lengthy treatise on the subject, contrasting the principles of geocentrism with heliocentrism in the form of a discussion, which was titled *Dialogue on the Two Great World Systems*. Upon its publication, Galileo firmly believed that anyone who read it would immediately realize the accuracy of the Copernican system and reject any and all other models in favor of it.

This, however, was not to be. One of the characters in his book espoused the opinions of the Pope, thus drawing the ire of the Church. Galileo was soon summoned by the Inquisition under suspicion of heresy. By the end of his trial, he had been forced to recant his assertions regarding the heliocentric model and his *Dialogues* became another banned book. He was sentenced to house arrest and lived the rest of his life in isolation.

While Galileo's ability to impart further influence may have been limited at that point, there is evidence that the scientific world became greatly shaken by his works. In 1660, a Jesuit scholar named Athanasis Kircher published a book that presented a number of different models of the Universe, which he indicated were all possibilities. Among them were modified versions of the Ptolemaic model, where Mercury and Venus orbited the Sun while the Sun and everything else orbited Earth, modified versions of the Tychonic model, where Jupiter and Saturn orbited Earth outside the Sun's orbit around the Earth (with Mercury, Venus, and Mars orbiting the Sun), and the Copernican model. Including these with the more standard versions of the Ptolemaic and Tychonic model indicates that there was a tremendous amount of upheaval and discussion happening following Galileo's death.

As time ticked on, Galileo's observations and Kepler's calculations began to win over converts to the Copernican system and more astronomers began to adhere to a sun-centered model. While it eventually became widely accepted, the explanation of why planets orbit the Sun, why such motion of the Earth is imperceptible, and why objects still seem to be firmly attached (and even drawn) to a moving Earth would require yet another revolutionary mind—the great Isaac Newton.

CHAPTER FOUR

The cover to Isaac Newton's *Philosophiæ Naturalis Principia Mathematica (Mathematical Principles of Natural Science)*, sometimes simply referred to as Principia. This book is considered by most scientists to be one of the most important works in the history of science.

Isaac Newton / Public Domain

The Newtonian Revolution

Learning Objectives

In this chapter you will learn:
- » How philosophers in the Renaissance era viewed the Universe around them and their ability to understand it
- » The ways that Isaac Newton contributed to a radical transition in scientific thought
- » How Newton's description of gravity helped to substantiate Kepler's laws
- » Newton's three laws of motion and how he was able to describe the Universe with them

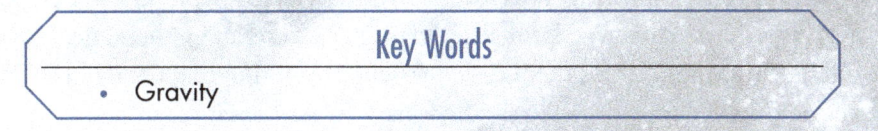

Key Words
- Gravity

4.1 Isaac Newton

Isaac Newton (1642–1727) was born in England less than a year after the death of Galileo. Newton (**Figure 4.1**) came into a world that was at a crossroads. On one hand, many individuals still believed in witchcraft and magic. To these people, the Universe was a mystical realm that could never be fully understood. On the other hand, the field of philosophy was approaching a dividing point where many of the world's great thinkers were beginning to resemble modern scientists. Such people were asking questions and proposing tests for their hypotheses rather than simply using rhetoric to support their positions. The world was a very interesting place.

Astronomers were gradually coming to accept the heliocentric model and the way that Kepler's laws of planetary motion served to describe the motion of planets in their orbits around the Sun. However, little was known about why these planets ought to move the way they do.

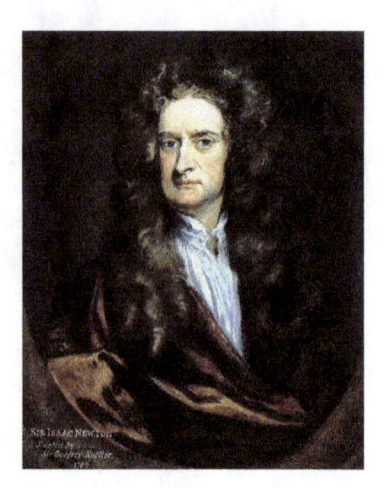

Figure 4.1. Sir Isaac Newton, considered by many to be one of the greatest scientists the world has ever seen.

Sir Godfrey Kneller / National Portrait Gallery, London / Public Domain

Kepler himself believed that the planets would move only if compelled to do so. In his view, then, the Sun was continually tugging on each planet, pulling it along like a stubborn mule. Other philosophers, such as René Descartes, believed that planets would normally move in a straight line but that the rotating Sun was creating a swirling vortex around it, thus forcing the planets to follow orbiting paths instead.

Instruments that existed to make measurements were not terribly accurate by modern standards. While clocks had been around for many centuries, their ability to measure fractions of a second was limited. This meant that experiments designed to study motions of objects were often not precise enough to distinguish between predictions of competing hypotheses. Indeed, even the concept of motion was still poorly defined—Aristotle had included such things as the ripening of an apple or a child growing as examples of bodies in motion.

As Isaac Newton grew and began his studies at Cambridge University, he became exposed to the work of Galileo and Pythagoras. Several decades earlier, Galileo had studied the subject of motion and had announced that all bodies on Earth seem to fall at the same rate—that is, their speed increased at the same rate regardless of how massive the object was. This concept of uniform acceleration proved intriguing to Newton, as he wondered why this should be the case.

At the point when Newton was preparing to tackle these questions, bubonic plague broke out all over England and Cambridge was forced to close. Retreating to his home, Newton dove headfirst into his studies. However, he realized that there were no mathematical tools available to him to fully address the concept of motion, particularly motion that changed instantaneously through acceleration. He was forced to develop his own mathematical methods to consider the problem and in doing so he invented what is now known as calculus—the mathematical study of how quantities change.

4.2 Universal gravitation

Having developed his methods of calculus, Newton was then able to mathematically study the forms of motion and what would produce such forms. Focused after (and perhaps inspired by) heated exchanges with Robert Hooke, another English mathematician and philosopher, Newton applied his mathematical prowess to the question of the attractive force of nature that might cause planets to follow elliptical orbits around the Sun.

In doing so he envisioned this force, referred to as gravity, as a "drawing power" intrinsic to all matter. He famously told of his inspiration for this concept after watching an apple fall from a tree and wondering why the apple should always fall vertically, rather than any other direction. Surmising that the Earth must draw the apple to itself, he correctly deduced that the apple must also be drawing the Earth to it—a force of mutual attraction. Mathematically, he demonstrated that this gravitational force should take on the form

$$F = G\frac{M \cdot m}{d^2}$$

where the masses M and m correspond to the two objects attracting one another, d is the distance between their centers, and G is called the gravitational constant and represents the magnitude of the force, in physical units.

Look closely at the equation above. Notice that it requires *two* masses in order to calculate the force between them. This indicates that gravity is not a force that an object feels due to the presence of another object, nor is it a force that one object exerts on another (as it is commonly described). Rather, it is a force that exists only between the two objects due to the presence of both objects. If the Universe contained only one object, while this object would possess a gravitational field related to its mass, there would be no way to define a gravitational force. In order to do this, there needs to be a second object. Furthermore, regardless of which object you assign as M and m, the force is the same. Gravity is a force *between* two objects—not simply a pulling of one object on another.

> **Gravity**
>
> The gravitational force is an attractive force between two objects with mass, where objects pull on each other with the same force, rather than simply one object pulling on the other.

Now look again, this time at the denominator. If you increase the separation distance d between the two objects, the force decreases. Make that distance larger and larger causes the force to get smaller and smaller. Interestingly, though, the force can never go to zero unless the distance reaches infinity. Even after plugging a huge number into the denominator, like one million, the force still has a value that is not zero. It's small, but not zero. This means that, technically speaking, there is no region in the Universe that is free from gravity. See **Tools of the Trade 4.1** for a mathematical example of Newton's law of universal gravitation.

Using his law of gravity, Isaac Newton was then able to show that there exist three fundamental laws of motion. While these were not entirely unknown at the time, Newton was able to show that they result naturally in a world where his law of gravity applied:

Newton's first law represents the idea of inertia, where an object will essentially continue doing whatever it's doing (moving or just sitting there) as long as nothing interferes with it. Newton's third law represents the idea of balanced forces in equilibrium, a principle that is fundamental to understanding whether an object will move under the influence of one or more forces or remain stationary. Engineers that construct bridges and elevators apply this principle so that bridges remain strong and elevators operate smoothly with minimal strain on the gears and motors.

Tools of the Trade 4.1

The Law of Universal Gravitation

Let's look at gravity a little more closely to understand what it reveals about the interaction between two objects. Suppose we choose our units for mass and distance such that we can ignore G; we will set it equal to 1. This leaves us with the equation for gravity as

$$F = \frac{M \cdot m}{d^2}$$

Now suppose we have a situation where there are two objects. The larger object has a mass of 4, and the smaller object has a mass of 2. They are separated by a distance of 4 units.

Let's set $M = 4$ and $m = 2$. Now we can use the equation above to calculate the force between the two objects:

$$F = \frac{M \cdot m}{d^2} = \frac{4 \cdot 2}{4^2} = \frac{8}{16} = \frac{1}{2}$$

Now, as a check, let's reverse the mass values. Suppose instead we say that $M = 2$ and $m = 4$. Plug these numbers in again and calculate the force:

$$F = \frac{M \cdot m}{d^2} = \frac{2 \cdot 4}{4^2} = \frac{1}{2}$$

The forces are the same, regardless of which object we assign to be M and m. What does this mean? This means that the gravitational force on the red object due to the blue object is identical to the force on the blue object due to the red object. Even though the red object is smaller in size and mass, the gravitational force from its perspective is identical.

In terms of Newton's observation of the apple falling to the Earth, the observation that the Earth fell to the apple is equally correct. Both objects fell toward each other, but the movement of the Earth toward the apple was imperceptible. Likewise, the Earth "pulls" on the Sun with the same gravitational force that the Sun pulls on the Earth. However, since the Sun is so much larger than the Earth, any motion it experiences is less noticeable in comparison to the effect on the Earth.

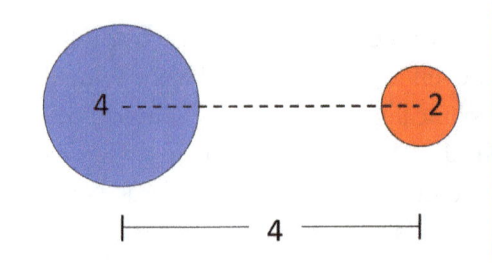

Figure 4.2.

Newton was able to show that this force of gravity applied to the Moon orbiting the Earth and all planets orbiting the Sun. Hence, it became known as the Law of Universal Gravitation. Being able to express it in mathematical form was a major success and Newton gained a reputation as a genius for his work.

> **First Law of Motion:** An object at rest remains at rest, and an object in motion remains in motion at constant velocity in a straight line, unless acted upon by an *outside force*.
>
> **Second Law of Motion:** The rate of change of an object's motion is directly related to the magnitude and direction of the force acting upon it. That is, $F = ma$.
>
> **Third Law of Motion:** For every action (force), there is an equal and opposite reaction. Forces of two bodies on each other are always equal in magnitude and opposite in direction.

One of Newton's other important achievements was to provide an impetus behind Kepler's laws of planetary motion. Demonstrating gravity as a force of nature, he was able to mathematically prove that all of Kepler's laws resulted from gravitational interactions between the Sun and each planet, even the occurrence of elliptical orbits. Additionally, he showed that Kepler's third law could be generalized to apply not only to planets orbiting the Sun but, in the appropriate units, also the Moon around the Earth, the Galilean moons orbiting Jupiter, and to newly discovered moons orbiting Saturn. In doing so, he provided a sweeping unification of physics on the Earth and physics throughout the Universe.

Figure 4.3 (a). Astronaut Bruce McCandless II, mission specialist, participates in an extra-vehicular activity (EVA) using a nitrogen-propelled hand-controlled manned maneuvering unit (MMU). He is performing this EVA without being tethered to the shuttle. (b) Astronaut Catherine Coleman floating in the Japanese Kibo laboratory on the International Space Station during a live NASA TV broadcast on Saint Patrick's Day 2011, while playing a tin whistle belonging to musician Paddy Maloney of The Chieftains.

(left) NASA / Public Domain; (right) NASA TV / Public Domain

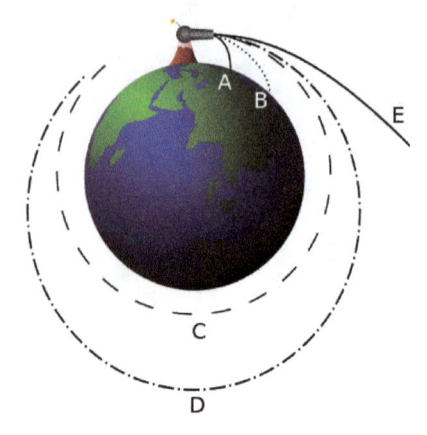

Figure 4.4. Newton's cannon demonstrates that when a cannonball is fired with a small velocity (a) it will soon fall to Earth due to the force of gravity. However, as the launch speed is increased, there comes a point when the cannonball, still falling under the force of gravity, effectively misses the Earth and remains at the same altitude from which it started (c). The cannonball is now in a circular orbit.

Copyright © 2007 by Brian Brondel / Wikimedia Commons / CC BY-SA 3.0

4.3 Additional contributions

In formulating a correct understanding of gravity, Newton was also able to explain the Earth's tides as a result of the differing gravitational pulls of the Moon and Sun on liquid water bound to a solid, rotating Earth. Using his generalization of Kepler's third law of planetary motion, masses could be determined for any planet that was known to have a moon; at the time, this included Earth, Jupiter, and Saturn. Astronomers determined that Jupiter and Saturn were very massive compared to the Earth.

One other revelation from Newton's work on gravity was a more accurate description of objects in orbit. We are accustomed to seeing astronauts floating within the Space Shuttle or International Space Station (**Figures 4.3a** and **4.3b**), and many people incorrectly assume that this means there is no gravity in space. We now understand that there is no place where gravity cannot reach. The fact that objects exist means they have gravitational influences that extend to vast distances. So how, then, do these astronauts seem to defy gravity?

Isaac Newton himself demonstrated that such objects in orbit are actually in a state of free fall. He used the example of a cannonball fired from a large cannon. **Figure 4.4** shows the cannon Newton envisioned, which fires a cannonball that falls under the force of gravity (a). As the ball's launch speed is increased, it travels farther while still falling, until it lands at a greater distance than the first ball (b). If the launch speed is chosen appropriately, the ball can travel so far forward that as gravity pulls it downward it effectively misses the Earth, remaining at the same altitude from which it was originally fired. This cannonball, in free fall under the force of gravity, continues in a circular orbit around the Earth (c). The launch speed can be increased to produce a more eccentric orbit (d) or even to allow the cannonball to escape Earth's gravity altogether (e).

Orbiting objects like the astronauts in **Figures 4.3a** and **4.3b**, the International Space Station, and even the Moon are therefore in free fall around the Earth. Their sideways speed is great enough to keep them from hitting the Earth as they fall. If you think about it, this even means that the Earth is in free fall around the Sun!

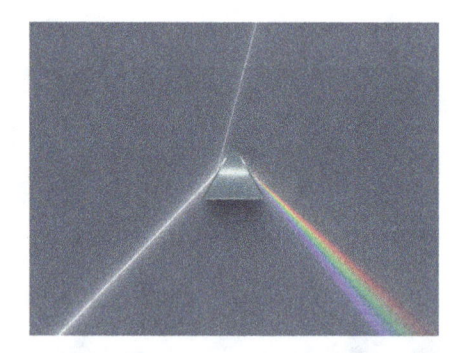

Figure 4.5. Using a glass prism, Isaac Newton discovered that a beam of white light can be broken up into different colors. The observation that white light is composed of all the colors of the rainbow led to tremendous insight about the spectrum and the nature of light.

Copyright © 2010 by User:Spigget / Wikimedia Commons / CC BY-SA 3.0

During his period of incredible intellectual accomplishment, Newton also set his mind to other matters of physics. He studied the nature of light, determining through experimentation that a beam of white light was actually composed of all the colors of the rainbow, seen in **Figure 4.5**. While it had been known up to this point that a prism could reveal such a rainbow, it had been unclear whether or not the prism actually produced the colors. This led to the discovery a century later by William Herschel of the first non-optical portion of the spectrum: infrared light.

While studying the nature of light, Newton also noticed that unlike sound waves, which can be bent (refracted) around obstructions, light seems to cast sharp shadows. This inability for light beams to be bent around obstructions suggested to him that light must be composed of particles. Other scientists, however, pointed out that light can be refracted when it passes through a denser medium (such as water or a glass prism). This suggested that light must be composed of waves. Competing models of light as particles and as waves were studied for several centuries before Albert Einstein, in the early 20th century, showed that light is *both* particle and wave. We will look at this wave–particle duality in greater detail in Chapter 6.

One final contribution that Newton made to the field of astronomy, which had a monumental impact on the ability to observe celestial objects, was his invention of a reflecting telescope that bears his name. The Newtonian reflecting telescope used a curved mirror to focus light, rather than a glass lens as Galileo's telescope had. Because mirrors were easier to make than glass lenses, and because they could be made larger and their weight could be more easily supported, the Newtonian telescope quickly took off and became the primary design used in all major telescopes for the next two centuries. We will look at the history of the telescope, and the designs of major telescopes, in Chapter 7.

Isaac Newton's professional life was exceedingly focused and productive. His contributions to the body of knowledge that had been established up to this point in history were arguably unrivaled. Following Newton's description of the gravitational force, a fundamental change occurred in the way that people viewed the Universe. It had been initially thought that all of space was a sort of fluid, which allowed an object like the Sun to gravitationally influence the Earth despite its great distance. Newtonian gravity introduced the idea of "action at a distance" and the Universe came to be viewed as a largely empty grid through which moved the planets, orbiting the Sun like clockwork. While

Figure 4.6. A replica of Newton's reflecting telescope. A curved mirror is placed on the opposite (left) end as the opening where the light enters. This mirror then directs the light up to a flat mirror near the opening, which redirects it to a focus through a hole, drilled in the side of the telescope tube. An eyepiece can then be placed there to observe the image.

Copyright © 2004 by Andrew Dunn / Wikimedia Commons / CC BY-SA 2.0

there remained a notion of a permeating substance (called the *ether*) that allowed light to be transmitted from the Sun to the Earth, this ether was not viewed as a fundamental structural element. The view of the Universe had been radically changed, and with it scientists had been given the mathematical tools (calculus) and observational instruments (telescopes) to take the next leap forward in astronomy. This next leap, however, would not come for another two hundred years.

CHAPTER FIVE

Albert Einstein, in 1921. Einstein's remarkable sequence of publications opened the door to the modern era in physics and astronomy and introduced a vastly new way of viewing the Universe.

Ferdinand Schmutzer / Austrian National Library / Public Domain

The Dawn of Modern Physics

Learning Objectives

In this chapter you will learn:
- » How Einstein demonstrated that the speed of light has a value that is constant and independent of one's state of motion
- » How Einstein's views of the Universe differed from those of Isaac Newton
- » The equivalence of mass and energy
- » Details of special relativity that describe how high speed affects perception of time, size, and mass
- » The nature of how general relativity changed the way we think about the framework of the Universe

Key Words
- Photon
- Time dilation
- Length contraction
- Spacetime

5.1 A shift in perspective

By the end of the 1800s, astronomy and physics had come a long way since the Newtonian revolution. Large telescopes had been constructed capable of observing extremely faint objects, imaging techniques were being developed, and theoreticians were hard at work trying to explain the phenomena of the solar system and stars. As scientists unveiled broader and deeper understandings of the world around them, several interesting experimental and mathematical questions arose that needed to be explained:

1. What is the nature of atoms and molecules? Are they real, or merely a convenient mathematical construct to describe observations?
2. What is the nature of light? What is it composed of? Does it require a medium to travel through, as sound waves require air?
3. Why are some substances radioactive, emitting particles seemingly at random?
4. How does the Sun produce dark lines in the light spectrum it emits?

It would take a beautiful mind to tie all of these questions together.

Albert Einstein (1879–1955) showed intellectual curiosity at an early age. When he was shown a compass at age 5, he marveled at how the small metal needle would orient itself to point north regardless of how its case was turned or jostled. What invisible force could be exerting itself upon this device?

As he grew, the young Einstein (**Figure 5.1**) demonstrated tremendous skill in physics and mathematics, perhaps motivated by his interest in the subjects. He entered college to study physics, graduating in 1900. However, unable to find a job teaching at a university, he picked up a job working as a clerk in a Swiss patent office in Bern. It was while working there that he continued his studies toward a doctorate degree, which was awarded to him in 1905. His dissertation, exploring the nature of molecules, led him to one of his first publications that would make him famous. In fact, his publication of four groundbreaking papers in 1905 eventually led many historians to refer to that year as his "annus mirabilis" (miracle year).

Figure 5.1. Albert Einstein as a child. Even at a young age, Einstein displayed the curiosity of a budding scientist.
Wikimedia Commons / Public Domain

Einstein's first paper explored a phenomenon called the photoelectric effect (**Figure 5.2**). In this process, light is absorbed by a substance and electrons are emitted. While the nature of atoms was not yet fully understood, Einstein was able to show that the reason electrons can be emitted in such a way is that light behaves like "energy packets": bundles of energy that can be absorbed by a substance. When certain values of this energy are absorbed by electrons, this causes them to be ejected from their initial states. The observation of such a phenomenon did not fit with a description of light as a wave, as had been accepted over the previous century. The notion and demonstration of light as a particle, a *photon*, led Albert Einstein to be awarded the Nobel Prize in 1921. This heavily promoted the newly developing ideas of quantum mechanics, which considered the behaviors and interactions of particles on extremely tiny scales.

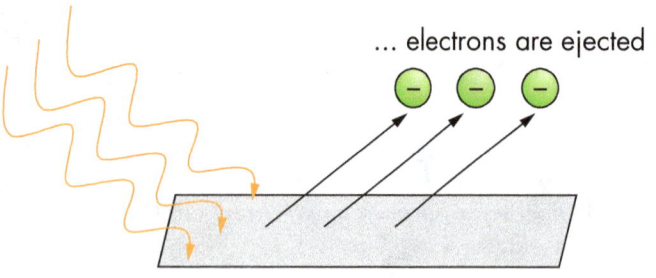

Einstein's second paper of 1905 centered on Brownian motion, the erratic movement of tiny particles suspended in a liquid. Einstein was able to show that the motion of these particles comes from their being continually bombarded

Figure 5.2. The photoelectric effect

by molecules of the liquid. By showing that because of these collisions particles were unable to remain still, he was able to prove the reality of molecules—particles too tiny to be seen with the human eye.

> **Photon**
> A particle of light, often described as a bundle of energy, which can interact with other particles.

Taken together, these two papers published by Albert Einstein during the first half of 1905 would have constituted a wildly successful career for any scientist. By showing that molecules exist, Einstein opened the door for the exploration of matter at its most fundamental levels. By showing that light behaves like a wave in some circumstances, which had been previously studied, and as a particle in other circumstances, Einstein began to pull the curtain away to reveal the mysterious nature of light. It would later be discovered that even matter particles like electrons behave as waves under certain conditions.

A third publication from Einstein in 1905 demonstrated the equivalence of mass and energy via the famous equation $E = mc^2$. Energy and mass are in essence the two sides of the same coin. According to this principle, an object's mass is considered part of its overall energy content and this energy can be liberated under certain processes, such as nuclear fusion. It also gave an indication that while mass can be converted into energy, energy can be converted into mass. This idea would prove pivotal in describing the development of matter in the early Universe.

Einstein's work on the quantum nature of light and matter fundamentally impacted the field of physics on small sizes. Perhaps his greatest act, initiated with a fourth paper in 1905, would impact the field on larger scales.

5.2 The special theory of relativity

5.2.1 What is Relativity?

Having studied the newly published ideas and equations of electricity and magnetism from James Clerk Maxwell, Einstein was struck by an interesting point: the speed of light was defined by the product of two fundamental physical constants of nature. Since the observed speed of an object is partly determined by the motion of the observer, it struck Einstein as odd that the speed of light should have no similar condition.

Centuries before him, Galileo had also considered the subject of motion. He pointed out that if a person was in an enclosed, windowless room that was moving at a constant speed, then that person would have no way of knowing they were in motion. Any experiment they performed in their enclosed room would turn out exactly the same as it would if it were performed on the ground, stationary. This was the first articulation of a principle of relativity, and also helped explain why Earth-bound humans cannot sense the Earth's motion in orbit around the Sun.

Consider, for example, an American football quarterback who is looking to throw a pass. Suppose this quarterback can throw a football at a speed of 50 miles per hour while standing still. If the quarterback takes off and runs forward at a speed of 15 miles per hour while throwing the ball, a stationary observer in the audience would see the football fly through the air at 50 + 15 = 65 miles per hour.

Similarly, if the receiver is standing still, then he too would observe the ball approaching him at 65 miles per hour. On the other hand, if he is running downfield at a speed of 15 miles per hour as the ball is coming toward him, then he would observe the ball approaching him at a speed of 65 − 15 = 50 miles per hour. In summary, the observed speed of the football depends on the motion of the observer, and in this case the thrower as well. (**Figure 5.3**)

Figure 5.3. The football's total speed in the air is the sum of its thrown speed (50 mph) and the quarterback's running speed (15 mph). If the receiver is stationary (a), that is the speed at which he sees the ball approach him. If the receiver is running away from the ball as it approaches (b), then he sees the balls approach speed reduced by his own running speed. The speed of the ball as perceived by the viewer—the receiver—depends on the viewer's own speed, so the quarterback and receiver would disagree on how fast the ball seemed to be moving.

With light, however, things are different. The speed of light is now known to be approximately 300,000 kilometers per second, or about 670 *million* miles per hour, and is often referred to with the letter *c*. If a person could measure the speed of the photons coming out of a flashlight she was holding, she would measure their speed to be equal to *c*.

Suppose Sammy is floating out in space in absolute darkness. Lacking any reference points and feeling no forces, Sammy is free to assume that he is stationary. As he hangs suspended in space, he notices a figure approaching in the distance at a constant speed. As the figure gets closer, he recognizes that it is Sandy, another astronaut. Sandy floats past him, they wave to each other simultaneously, and Sandy floats away. From Sammy's perspective, Sandy is the one in motion.

From Sandy's perspective (called a *reference frame*), she is stationary. She

Figure 5.4. Sammy views Sandy's speed relative to him to be $v = 0.75c$. Despite this relative motion, as Sandy shines her flashlight in front of her, both would agree that the light beam is traveling through space at the same speed—the speed of light, *c*.

hangs suspended in space and gradually sees Sammy approaching at a constant speed. As he passes her, they wave to each other simultaneously and he continues on his way. From Sandy's viewpoint, Sammy is the one who is in motion.

According to the principle of relativity, both perspectives are equally valid. In the absence of any reference points, both assume they are stationary and that the other person is moving. There is, in this situation, no way to say that either perspective is wrong. This illustrates the concept of relative motion. Motion can be defined only with respect to a reference point. Since this example lacks reference points, there is no way to define who is in motion.

Similarly, both astronauts are experiencing no forces that are pushing or pulling them along. This places them both in what is called an *inertial reference frame*. Inertial reference frames represent constant velocity motion and lack any sort of acceleration that produces a force. The absence of forces also prevents either astronaut from definitively acknowledging that they are moving.

Suppose Sammy was observing Sandy flying through space at a very high speed, say 75% of the speed of light ($0.75c$), and while doing so she turned on a flashlight and pointed it forward, shown in **Figure 5.4**. In Sandy's perspective if she was traveling at a constant speed then she

Sandy's Prespective

Sammy's Prespective

Figure 5.5. From Sandy's perspective (a), she is at rest and Sammy is moving away from her at $0.75c$, while she sees her beam moving away from her at c. Since she does not perceive her own motion, though she may have been told she is moving, she will always see the beam moving away from her at c and therefore cannot ever hope to fly fast enough to catch it. According to Sammy (b), Sandy is moving away from him and he also sees her flashlight beam traveling at c. Both agree that the light is outpacing Sandy's motion, so for this to always remain true then regardless of Sandy's speed, she will never travel fast enough to catch the light beam.

should perceive herself to be stationary, as Galileo explained, with all of space traveling past her at a speed of $0.75c$. If Sandy can state that she is stationary, then when turning on a flashlight and shining it out in front of her, she should expect to see nothing surprising. That is, she should observe the light traveling away from her at a speed equal to c. She would report this to Sammy.

What Einstein realized, as noted earlier, was that the speed of light does not seem to depend on the motion of the observer or the emitter. Sammy would actually also measure the speed of the photons from Sandy's flashlight to be moving at a speed equal to c, rather than $1c + 0.75c = 1.75c$ as Galileo would have predicted. Both would observe the light traveling at c.

Sammy would then claim that Sandy should be seeing the light recede from her at only $1c - 0.75c = 0.25c$, since he sees her moving in the same direction as the beam but slightly slower than it. This might seem to contradict what Sandy has just reported, until she points out that she is the one who is stationary and Sammy is actually moving away from her at $0.75c$, giving him a mere illusion that this is the case, where instead his motion is impacting his view of her. Both agree that the light beam is moving at a speed equal to c, but since they disagree on who is moving, they also disagree on what the other's relative motion with respect to the beam ought to be.

Now, the next natural scenario that comes to mind is this: suppose Sammy sees Sandy fly by at an astoundingly high speed. Can she ever reach the speed of light? And if so, what would she observe? This situation is illustrated in **Figure 5.5**.

From Sandy's perspective, regardless of what speed she is going, she can still claim she is stationary, with all of space flying past her at high speed. This means that when she turns on her flashlight she will see the beam travel out in front of her. Even if she happens to know that she is the one traveling at high velocity, as long as the velocity remains constant then she remains in an inertial reference frame. Seeing the light beam traveling away from her at c means she must be traveling at a speed less than c. From her perspective, she can never reach the speed of light.

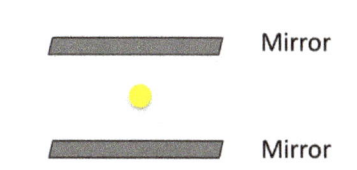

Figure 5.6. A diagram of a simple clock that ticks off the passing of time as the photon (yellow circle) bounces off each mirror above and below it.

As Sammy watches, he shouldn't be able to make a different claim. If Sandy sees the beam moving away from her, Sammy should too. He sees her beam moving at c, and is therefore correct in stating that Sandy must be moving at a speed less than c. This means that even from any other reference frame, Sandy cannot reach the speed of light. It just isn't possible.

Why should photons behave differently than other moving objects? And what else does this imply? To investigate further, Einstein proceeded with two assumptions based on this: 1) the laws of physics behave the same for any observer in an inertial reference frame; this results in 2) the speed of light is observed to be a constant c regardless of the motion of the observer or the emitter.

5.2.2 Time Dilation

In order to consider the effects of relativity and the speed of light, suppose our two astronauts each build themselves a clock. Fundamentally, a clock is any mechanism that can "tick" regularly in a constant, uninterrupted way. In order to produce a clock that the other could see, each astronaut builds a simple clock that consists of two mirrors facing each other, one above the other, with a photon that bounces from one mirror to the other. A schematic of such a clock is shown in **Figure 5.6**.

Sammy finishes building his clock and admires his handiwork while out in space once again. He watches the photon travel at the speed of light and bounce off each mirror, ticking off the passage of time. Looking up, he notices Sandy now approaching, also holding a clock that is identical to his own. He sees that the mirrors are separated by the same distance and that Sandy's clock also has a photon bouncing between the mirrors.

As Sandy's clock passes by, depicted in sequence in **Figure 5.7**, he compares the amount of time it takes for her photon to make one trip up and down (t' ["t-prime"] in the diagram) to the amount of time it takes for his photon to make the same trip (t in the diagram). His photon travels in its stationary clock up to the mirror in one tick (**Fig. 5.7b**) and down on the next

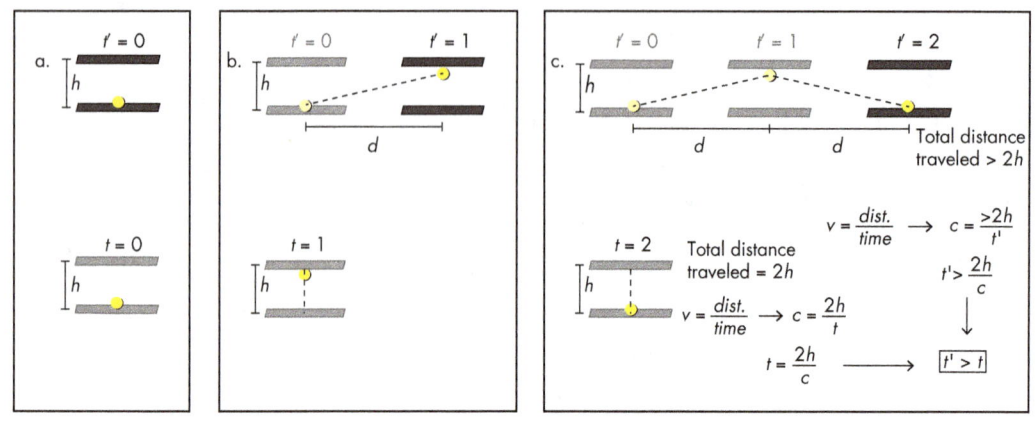

Figure 5.7. Tracing the path of Sandy's photon clock (top in each frame) and Sammy's photon clock (bottom in each frame). Viewing Sammy's clock as stationary, his photon travels a distance of *2h* as it bounces up and down. Viewing Sandy's clock as moving, her photon travels a distance greater than *2h* (dashed line) as it bounces up and down. Since Sammy sees both moving at the speed of light *c*, but watches Sandy's photon travel a greater distance at that speed, Sammy concludes that it takes longer for Sandy's photon to travel up and down once. Therefore, he believes that Sandy's moving clock is ticking *slower* than his stationary clock.

(**Fig. 5.7c**), traversing a distance $2h$ in the process as it moves at the speed of light c. The time it took to make this journey, then, is given by $t = 2h/c$.

As he watches Sandy's photon move up toward the top mirror, though, he notices that it has to also move sideways to keep up with the moving mirror*. This means that by the time it gets to the top mirror, the distance it has traveled is greater than h (**Figure 5.7b**), meaning that by the time it reaches the bottom mirror again it has traveled a distance greater than $2h$. Since the speeds of both photons are the same, this means that the amount of time Sammy says it took for Sandy's photon to go up and down once was $t' > 2h/c$—a longer time span than Sammy's clock. According to Sammy, then, Sandy's clock is running more slowly than his.

Sandy, on the other hand, can say the same thing about Sammy's clock. In her reference frame she is stationary and Sammy is in motion. She observes his clock running more slowly than her own for the same reason.

This observation is not just a trick of clocks and photons, though. It represents an intrinsic slowing of the perceived passage of time in a moving reference frame according to a stationary observer. Sammy not only sees Sandy's clock running more slowly as she passes by him, but he also sees her blinking more slowly than he does.

> **Time dilation**
> The effect of special relativity where a moving clock appears to run more slowly than a stationary clock.

---

* Both agree that Sandy's photon hits both mirrors. Since they are in motion at constant velocity, both astronauts can take the position that they are stationary and therefore expect their experiment to behave as if no motion is occurring. It would be very strange to Sandy if her photon missed the mirror simply because Sammy was watching it!

Her wave to him seems to move a bit more slowly than his. If he could see her cells functioning, he would see them dividing more slowly than his own cells. According to Sammy, who is stationary in his reference frame, Sandy is actually *aging* more slowly. The passage of time has slowed down in Sandy's moving reference frame. This effect is called *time dilation*.

Of course, Sandy could argue until the cows come home the very same point about Sammy. In her point of view, she is stationary and Sammy is moving. His clock is running slow, his body is aging more slowly, and time has slowed down for him. Without any reference points or forces to indicate which one is truly in motion, they are both arguing equally valid points. Neither one is wrong!

The conscientious student will ask, "Is this just an apparent effect, as a result of the motion? Once the motion is stopped, will the two clocks actually sync back up?" The answer is no: the clocks will not sync back up, and by causing the moving reference frame to stop the correct perspective will be revealed.

Time dilation has been experimentally demonstrated time and again and is a very real effect. Here are a few examples:

- Particles called muons are created when cosmic rays (high-energy particles flying through space) collide with particles in Earth's upper atmosphere, at an altitude of some 10–12 km. These muons have an average lifetime of approximately 2×10^{-6} seconds before they become unstable and decay into other particles, a lifetime that has been measured experimentally. Even if these muons were somehow created moving at the speed of light, they could travel only about 600 meters before decaying away. However, muons from the upper atmosphere are detected here on the ground. This can be explained only by the "internal clock" of a muon being time dilated when viewed from our reference frame, allowing it to live long enough to travel 10 km from the upper atmosphere to the ground. According to the muon, it still lives for only 2×10^{-6} seconds.
- In 1971, two atomic clocks were placed aboard commercial jets and flown around the Earth, one to the east and one to the west. The idea was that a reference clock on the Earth's surface was moving with respect to the Earth's location in space as the Earth rotated. This moving reference clock should experience a small amount of time dilation. If a plane could fly west, opposite the Earth's rotation, then it could establish a clock that was moving through space more slowly than the reference clock, thus experiencing less time dilation. Hence, it would tick a bit faster than the reference clock. The clock flying to the east, moving with the Earth's rotation, would be moving more quickly through space than the reference clock, marking the passage of time to be slightly more dilated than the reference clock. This clock would tick a bit slower and fall behind. When the jets landed, the westbound clock had indeed fallen behind by nearly 60 nanoseconds, while the eastbound clock had gained nearly 300 nanoseconds. This experiment was repeated

in 2005 with greater precision and the results once again supported the effect of time dilation predicted by special relativity.

- GPS satellites are programmed to take into account relativistic time dilation. Because they are orbiting Earth at such high speeds, their internal clocks should experience a certain amount of time dilation. When relativistic corrections are programmed in, their clocks run as if they were located on the Earth's surface.

5.2.3 Length Contraction

Suppose Sandy, again traveling at a speed of $0.75c$ with respect to Sammy, is heading away from Earth toward a star 12 light-years away. Sammy remains on Earth to watch her journey with his telescope. Moving at this speed, Sammy calculates that in 16 years she will have reached the star. However, he knows that her moving clock is running more slowly than his, so while he will see 16 years pass before she gets there, Sandy will experience fewer years passing before she arrives.

Approximately 10.5 years after she departs from Earth, Sandy arrives at her destination star. "But wait," she wonders, "if the star was 12 light-years away and I was moving at less than the speed of light, how did I get here in less than 12 years?"

> **Length contraction**
>
> The effect of special relativity where a moving object appears to be shortened along its direction of motion.

The solution to Sandy's dilemma is a phenomenon called *length contraction*. In Sandy's reference frame, she is stationary and all of space is flying by her at $0.75c$. Special relativity predicts that not only do moving clocks tick more slowly, but moving objects get shortened in the direction of their motion. This means that if space were moving past Sandy at $0.75c$, the length (distance) between her and the destination star was shortened to less than 12 light-years to the point where it only took 10.5 years to traverse the distance. The same is true for the muon mentioned above: from the muon's perspective, its clock is ticking normally but the distance from the upper atmosphere to the Earth's surface has been length contracted to the point that it can make the journey in 2×10^{-6} seconds before decaying away.

Let's look at another example to see how this works. Suppose Sammy is out in space, wearing a wristwatch and holding a yardstick. As he hangs suspended in space, he sees Sandy fly by at a relative velocity of $0.75c$, also wearing a wristwatch and holding her own yardstick. Sammy decides he is going to try to try to measure the length of her yardstick by timing how long it takes Sandy's yardstick to pass him by. If he knows her velocity (which has units of m/s) and measures the time that elapses (which has units of s), then he can calculate the length of her yardstick. You can understand this by watching what units result in multiplying velocity by time:

$$\frac{m}{s} \cdot s = \frac{m}{\not{s}} \cdot \not{s} = m$$

This multiplication produces units of meters, which is a length unit.

Sammy starts his timer when the leading end of her yardstick passes him, and stops it when the trailing end passes him, measuring an elapsed time t'. Sammy notices that Sandy is doing the same thing, measuring the length of Sammy's yardstick by counting off a time t. From Sammy's perspective, Sandy's clock is running a bit slower, meaning that the time t she measures is slightly less than the time t' that Sammy measures. Looking at the equation above, if the time (in s) is lower, then the length (in m) produced by the calculation is going to be lower too. This means that Sandy, who can claim she is at rest and that Sammy is the one who is moving, will measure Sammy's yardstick to be shortened by a small amount. Since both reference frames are equally valid in this series of examples, Sammy will also measure Sandy's yardstick, which is moving with respect to him in his view, to be shortened. The moving yardstick is *length contracted* in the direction of its motion.

Reread the preceding paragraph and consider it carefully. Both people perceive their own clock to be moving "normally" and that the other's clock, which is moving, is ticking more slowly. This results in a shorter time interval measured by the other's clock. A shorter time interval produces a calculated length that is shorter than what it "ought to be." Sammy will say that Sandy measured a length that was less than what he measured. Sandy will say that Sammy was moving, which means that his motion relative to her produced the shortened measurement. Relative motion produces length-contracted measurements.

5.3 Generalizing relativity

Einstein's formulation of special relativity was remarkable in its insight. One thing bothered him, however: it applied only to special situations, specifically when the motion involved a constant velocity. He was determined to provide a more general description of relativity under any circumstance, even when velocities are changing.

Continuing his employment at the patent office, Einstein spent the next 10 years of his life obsessing over countless thought experiments and mathematical derivations. Two years into this endeavor, he had what he called his "happiest thought." The revelation came to him while considering the nature of acceleration.

Under normal circumstances, we are accustomed to experiencing acceleration in tandem with a force. Driving a car, when we press the gas pedal the car accelerates, producing a force that pushes us back against our seat. When we make a turn in our vehicle, even if our speed doesn't change we still experience a force pushing us outward (either left or right) away from the center of curvature (**Figure 5.8**). A change in direction, even at constant velocity, still

Figure 5.8. Moving at a constant velocity, when your direction changes you experience a force due to the inertia of your initial motion. Similarly, hitting the brake causes a change in velocity, even if your direction remains the same, again producing a force due to your inertia. Experiencing these types of forces is a signal that you are accelerating (slowing down is a negative acceleration, or an acceleration directed in the opposite direction of one's motion).

Figure 5.9. According to Einstein's equivalence principle, a person accelerating in a windowless spaceship (bottom) would experience all of the same effects as a person at rest on the Earth's surface who is feeling the force due to gravity.

Copyright © 2007 by User:Mapos / Wikimedia Commons / CC BY-SA 3.0

constitutes an acceleration because velocity and acceleration are what are known as vector quantities, values that have both magnitude and direction. When the magnitude *or* direction of an object's velocity changes, the object is accelerating.

Try an experiment: with your left hand, hold out a sheet of paper parallel to the floor at eye level. Place your right hand on top of the sheet of paper. Release the paper with your left hand while quickly pushing the paper down with your right hand. By accelerating the paper in this way, it experiences a force backward against your hand, making it seem to stick to your hand. As we have experienced, then, an acceleration produces a force.

Isaac Newton had expressed this relationship as well, in his second law of motion. The equation $F = ma$ means that a force F produces an acceleration a, and vice versa. This led Einstein to his revelation. When objects fall under the force of gravity, they accelerate. What if a person could accelerate through empty space in such a way as to experience a force of the same magnitude as gravity, therefore feeling as if they are being held to the floor of their spaceship despite the lack of an actual gravitational force?

The ultimate conclusion he arrived at is called the *principle of equivalence*: the effects of acceleration are equivalent to the effects of gravity. **Figure 5.9** illustrates this idea. The experience of a person in a completely enclosed spaceship, accelerating at 9.8 m/s² (known as Earth's acceleration due to gravity), will be indistinguishable from the experience of a person on Earth—objects would fall downward in the same manner and the person would feel their weight in the same way.

This led Einstein to his next revelation. When Newton formulated gravity, the nature of this force was completely unknown. All that was known was that there exists this unseen force associated with massive objects that seemed to act over vast distances. Einstein realized that in order for gravity to produce an acceleration, there must be more to space than meets the eye.

When the sled is at the top of the hill, it has a high potential energy and low kinetic energy.

As the sled slides downward, its potential energy is converted into kinetic energy and it accelerates down the slope.

Figure 5.10. At a high position, the sled has high potential energy (a). As the sled moves downward under the force of gravity, its potential energy decreases and is converted into kinetic energy (b).

Suppose you held a ball in your hand out in front of you. Although this ball isn't moving, it has what is called potential energy (energy of position). This means that relative to the ground, the ball possesses a certain amount of energy that can be released and converted into kinetic energy (energy of motion) as it falls and accelerates. Similarly, when a child climbs into a sled at the top of a snow-covered hill, this child is at a point of greater potential, and as the child slides down the hill this energy is converted into kinetic energy. High potential energy and low (zero) kinetic energy at the top of the hill, low potential energy and high kinetic energy near the bottom. The result of this conversion of energy is acceleration down the slope of the hill, shown in **Figure 5.10**.

Einstein recognized that the force of gravity might simply be an acceleration along a similar slope in space. Physicists often consider a quantity called gravitational potential energy, where an object high above the Earth's surface has high gravitational potential energy and an object on the Earth's surface has low gravitational potential energy. We can think of this as being analogous to high and low points on the slope in **Figure 5.10**. So let's picture what that slope would "look like."

Imagine stretching out a bed sheet with a friend so that it is held aloft and flat. Now another friend places a marble on the sheet. This marble will produce a curved indentation in the sheet around it. If you dropped a pea onto the sheet, the pea would experience this curvature and

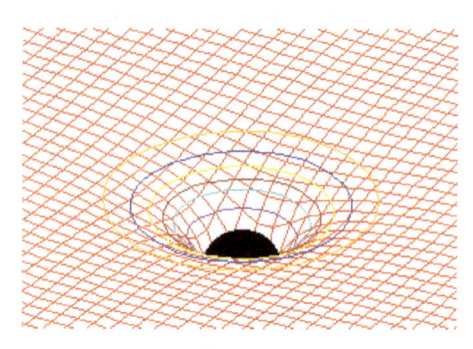

Figure 5.11. A depiction of how a massive object bends space around it, producing areas of lower gravitational potential energy that objects will tend toward.

Copyright © 2004 by User:Superborsuk / Wikimedia Commons / CC BY-SA 3.0

roll toward the marble as if the marble were pulling the pea toward it. Of course, it isn't that the marble is somehow "pulling" the pea toward it; the pea is simply responding to the curvature it is experiencing—the slope around it. It will tend to move from its location at high potential energy to the point of lowest potential energy, accelerating as it goes. If you replaced the marble with a baseball, the effect would be even more pronounced.

Einstein envisioned the Universe as being similar to the outstretched sheet. A massive object like the Earth produces a curvature in this "fabric" that causes smaller objects to fall inward toward it (**Figure 5.11**). This manifests itself as a gravitational force. The Sun, being even more massive than the Earth, causes a greater amount of distortion on a much larger scale.

It is important to realize when looking at **Figure 5.11** that this is only a two-dimensional representation. In reality, this distortion is present in all three dimensions (up-down, left-right, forward-backward), so you can try to picture rotating the image in **Figure 5.11** in every direction and it would look like that. This is very difficult to draw!

In order to completely and correctly represent the Universe, however, Einstein had to include time in this model as well. You have already seen that the passing of time is impacted by motion, and Einstein realized that the passage of time is also impacted by acceleration—an effect called gravitational time dilation. Deeper inside the curvature shown in **Figure 5.11**, time actually passes more slowly than outside of this curvature. Since time is therefore subject to the presence of a massive object, it behaves like the other three spatial dimensions as well. He was therefore led to conclude that time is also a dimension, the fourth dimension in what he referred to as *spacetime*.

General relativity's characterization of spacetime as being a flexible "fabric" that was subject to the presence of massive objects was a revolutionary way of thinking. Instead of there existing an absolute grid of space, where objects simply moved through it as time ticked along, the Universe became a dynamic place where objects moved along space, not through it, and were able to influence the passage of time as well. Gravity morphed from being a mysterious force acting over large distances into simply the result of two objects responding to the way each object curved spacetime around it. The Earth, being more massive than the Moon, produces a greater amount of curvature in spacetime. While the Moon also distorts spacetime, it does so

Spacetime:

The combination of one time dimension and three spatial dimensions into a four-dimensional Universe in which we live.

to a lesser degree and this distortion does not impact the Earth nearly as much as the Earth's spacetime distortion impacts the Moon. Thus, the Moon appears to orbit around the Earth.

Things became much clearer and much more interesting at the same time. One prediction of this new way of describing the Universe was that light should be obligated to follow any curvature it experiences as well. In Newton's Universe, light traveled in straight paths, unaffected by the presence of objects unless one got in the way. In Einstein's Universe, curvature was everywhere and light traveled along the fabric of spacetime, tracing out this curvature. If one could observe a distant light source close in the sky to a more nearby massive object, general relativity predicted that it would be possible to measure the light path's deviation as it was forced to bend around the nearby object. This effect came to be known as gravitational lensing.

Figure 5.12 illustrates this concept. Light from a distant star is being emitted in all directions, but one of those beams happens to be traveling in the general direction of Sun. As it passes by the Sun, the curvature of space due to the Sun's presence forces the light beam's path to be altered, redirecting it toward Earth. As the observer on Earth, we see the apparent position of that star (dashed line) in a different spot than it would be if the Sun were not there.

This prediction was testable: during a solar eclipse the Moon passes between the Earth and the Sun, briefly blocking out its light and darkening the sky. Scientists realized that if they could measure the apparent positions of stars close in the sky to the Sun, they could

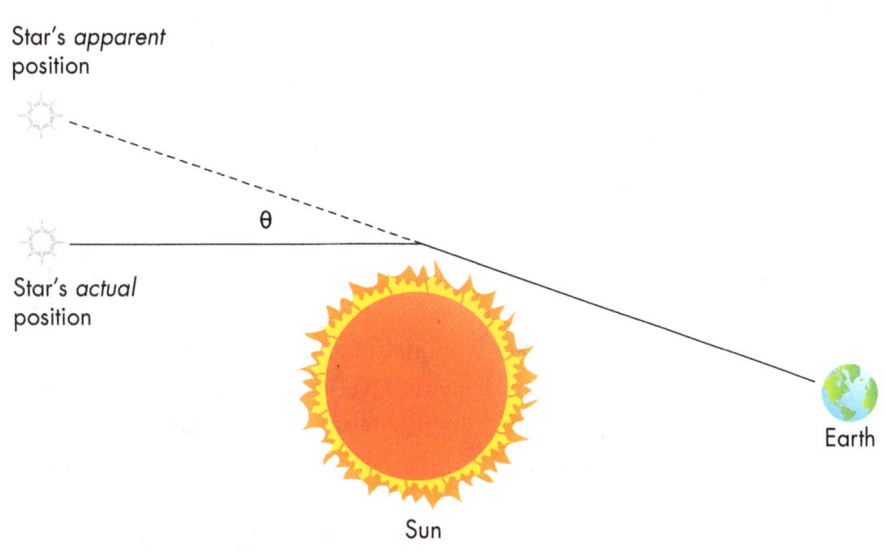

Figure 5.12. Gravitational lensing occurs when light from a distant object, such as the star shown above, passes close by a massive object (the Sun) that is closer to the observer. As the light is forced to follow the curvature of space, its path is deviated so that the observer sees it in an apparent position that differs from where it would appear if the massive object were not there. The angle of this difference in positions, θ, is predicted by general relativity and can be measured.

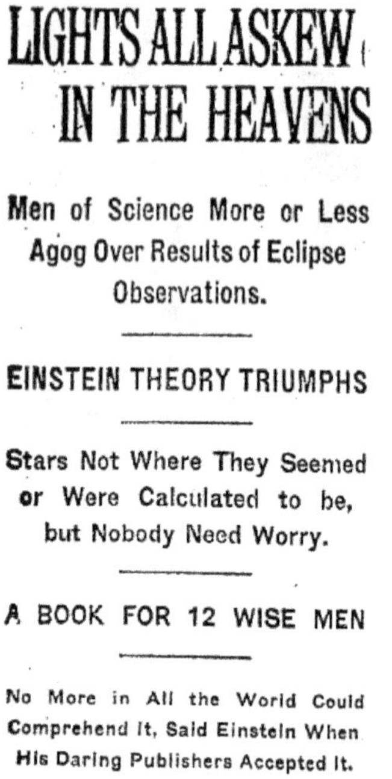
LIGHTS ALL ASKEW IN THE HEAVENS

Men of Science More or Less Agog Over Results of Eclipse Observations.

EINSTEIN THEORY TRIUMPHS

Stars Not Where They Seemed or Were Calculated to be, but Nobody Need Worry.

A BOOK FOR 12 WISE MEN

No More in All the World Could Comprehend It, Said Einstein When His Daring Publishers Accepted It.

Figure 5.13. A *New York Times* headline from Nov. 10, 1919 announcing the confirmation of Einstein's predictions. General relativity became widely accepted following the publication of Eddington's results.

The New York Times / Public Domain

compare those positions with the known actual positions and compare this difference (if it indeed existed) to the prediction of general relativity.

It just so happened that a solar eclipse in 1919 provided this opportunity. Sir Arthur Eddington traveled to the island of Principe, off the coast of Africa and photographed the solar eclipse. Using the images, he was able to measure the angle of deflection—the difference between where the star actually is and where it appeared during the eclipse—and showed that it agreed with the predictions of general relativity.

With this validation of what had been up to that point an extremely controversial hypothesis, Einstein gained worldwide fame (**Figure 5.13**).

5.4 Defining a new normal

The effects and predictions of special and general relativity may seem downright weird—we certainly don't experience time dilation or length contraction under typical circumstances on Earth. Our idea of "normal" is what we usually experience—that space and time are absolutes. Anybody should measure anything to be the same, regardless of what is going on. This is the construct of the Universe that Isaac Newton and his contemporaries gave us; the Universe as an absolute grid that we simply move through while time plods along in a consistent way.

Einstein's work in relativity shows us that while Newton's description of the world is fine for everyday matters (the Newtonian equation for gravity is quite accurate for all but the most extreme cases), the Universe simply isn't actually as Newton described. The Newtonian model is a mere approximation, useful for the basics but unable to accurately describe nature when tested at the extremes. Whether we are considering relatively humdrum phenomena like orbiting satellites and launched cannonballs, or the more fascinating behaviors of black holes, galaxy clusters, and muons, relativity provides a more correct explanation of the physics involved than does Newton's model.

Suppose Isaac lives in a cold environment where he never has to lift more weight than a glass of water. Lifting that glass of water is a very regular activity and the physical exertion behind it is "normal." One day, his friend Albert gives him, as a gift, a dumbbell with 30 pounds of weight on it. Upon lifting this 30-pound dumbbell several times, Isaac discovers that his body has started to secrete a liquid through his skin. "What is this?" he asks. Albert tells him that he has pushed his body a bit past what it had considered normal, and he is now experiencing something new: sweat. While this sweat might seem unusual to him, it is a very natural phenomenon. However, Isaac is only able to experience this "sweating" when he goes beyond a normal experience.

> The Universe we experience on a daily basis is only a drop in the bucket on astronomical scales.

Special and general relativity are very similar to this scenario. We are accustomed to moving at a relatively slow pace, interacting with objects that are not terribly massive, and experiencing size and distance scales that are quite small. It is only when we look beyond what is normal that we begin to see the true nature of the Universe. Looking beyond the small scales to which we are accustomed provides us with a glimpse into a Universe ruled by relativity.

In astronomy, there are many things that will take you beyond what you consider normal. To some, this is intimidating; to others, it is fascinating. As you continue to learn and continue reading, let the real nature of the Universe trigger your imagination. You will find out that everything we consider normal here on Earth is only a drop in the bucket on astronomical scales. In later chapters you will learn:

- The light our eyes can see is only a tiny fraction of the full spectrum
- The matter and energy we can see is only about 5% of all that actually exists
- The atoms that make up our body are mostly empty space
- The current size of the Universe compared to Earth is a billion times larger than the size of your body compared to an atom

The Universe truly is an amazing place!

CHAPTER SIX

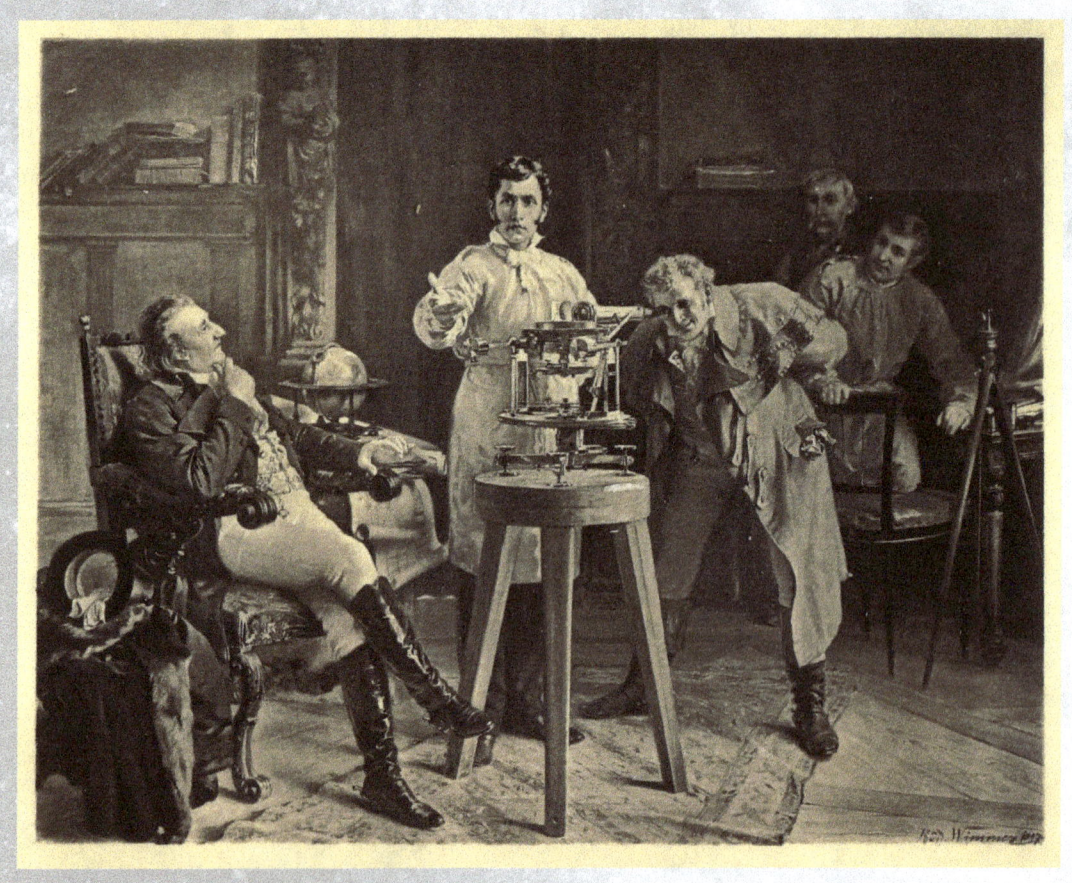

Joseph Fraunhofer is credited with having invented the spectroscope in 1814. His use of this spectroscope to study the Sun led him to discover dark absorption features in its spectrum, now called *Fraunhofer lines*.

Richard Wimmer / Public Domain

Light and Matter

Learning Objectives

In this chapter you will learn:

- » How the nature of the atom was determined, what its primary components are, and the principles of electron energy levels
- » The relationship between a photon's wavelength, frequency, and energy
- » The basic subdivisions of the electromagnetic spectrum
- » To name and describe the three types of observable spectra produced by the interaction of light and matter

Key Words

- Electromagnetic spectrum
- Resolving Power
- Absorption spectrum
- Emission spectrum
- Continuous spectrum
- Wien's Law
- Stefan-Boltzmann Law

6.1 The nature of light

Isaac Newton opened up the study of light through his work with glass prisms. By passing a beam of white light through a prism, a whole rainbow of colors appeared. This in itself was nothing new. What was not clear up to that point was whether the colors came from the light or from the prism. Newton devised the idea to place a second prism in front of the light from just one of the visible colors. As he watched one individual color enter the second prism, only the same color came out of it. This proved that the prism did not impart color into the light—the color was intrinsic to the light.

Figure 6.1. William Herschel, the man who introduced the world to non-optical radiation.

Lemuel Francis Abbott / National Portrait Gallery, London / Public Domain

One question went unconsidered, though: are the colors of light seen in this rainbow all there are to detect? This question was unwittingly answered in 1800 by William Herschel (**Figure 6.1**) as he studied the different colors of the spectrum.

Herschel's motivation was to learn about the different temperatures associated with each color. By placing a thermometer in each color of the spectrum, he noticed that the temperature rose as he moved it from violet to red. With the curiosity inherent to any good scientist, he wondered if this trend continued beyond the red light he could see. To his surprise, he discovered that the region just beyond the red was the hottest of all! Herschel had discovered what came to be known as infrared radiation.

It came as a shock to many in the scientific community to learn that there existed light beyond that which the human eye could detect. Over time, scientists realized that there exists a whole continuum of light, of which the visible portion is just a small fraction.

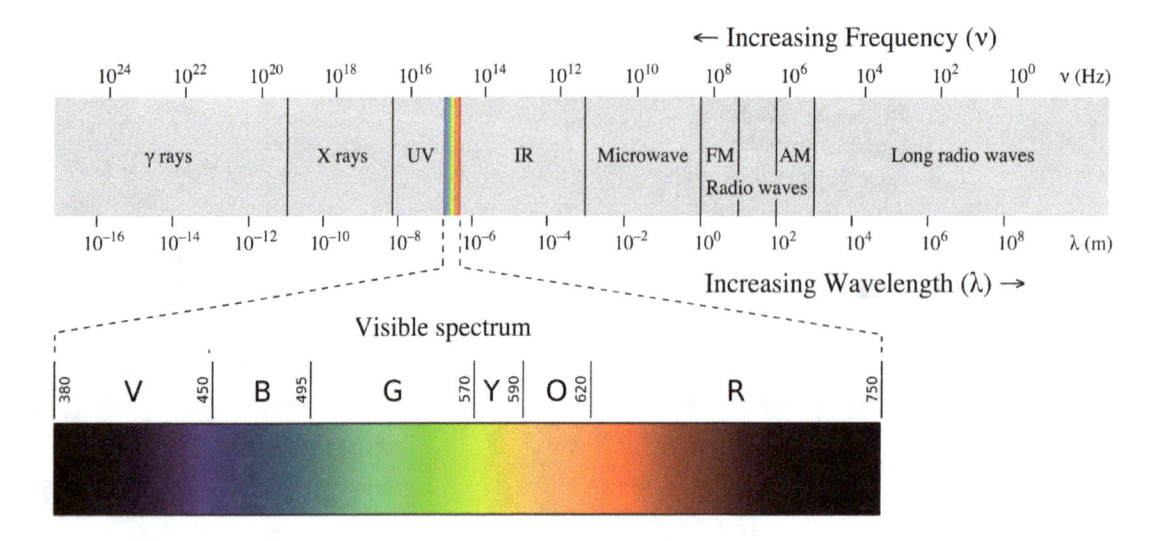

Figure 6.2. The full electromagnetic spectrum is divided into categories that are used to refer to each wavelength region. The visible spectrum, which our eyes are capable of detecting, is just a minuscule fraction of the full spectrum—which actually spans infinitely in both directions.

Copyright © 2013 by Philip Ronan / Wikimedia Commons / CC BY-SA 3.0

This continuum is now known as the *electromagnetic spectrum*, shown in **Figure 6.2**.

> **Electromagnetic spectrum**
> The full range of all possible wavelengths of light found in nature.

You learned in Chapter 5 that light is sometimes described as a particle called a photon. However, it can also be described as a wave. Light, as it turns out, is actually a product of vibrating electric and magnetic fields. These fields are all around us. When a charged particle vibrates, it produces an oscillation in the electric field it is creating. Physics tells us that when an electric field oscillates, it produces a magnetic field. These two fields oscillating together create electromagnetic waves, which manifest themselves as light.

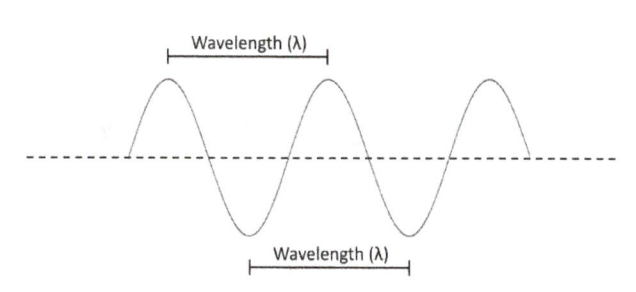

Figure 6.3. The wavelength (λ) can be defined as the distance from one peak to the next, or one trough to the next.

These waves can oscillate with a variety of wavelengths. The wavelength (λ "lambda") for any given wave is the distance from one point in the wave to the corresponding point in the next cycle. In the example shown in **Figure 6.3**, the wavelength can be the distance from one peak to the next or one trough to the next. Light waves can range from extremely short wavelengths to extremely long wavelengths.

When we measure distances in everyday life, we typically use such units as inches, feet, or miles. Using the metric system, typical units are centimeters, meters, or kilometers. However, when measuring a light wave's wavelength, these units are often *much* too large. Wavelengths of visible light are usually measured in nanometers (nm). The numbers in the visible spectrum inset in **Figure 6.2** are in nanometers. For shorter wavelength light waves, values are often expressed in an even smaller unit called an Angstrom (Å).

The portion of the spectrum our eyes can see, referred to as visible light, is generally depicted near the center of the spectrum. This placement is more by convention than anything else, since the spectrum technically extends infinitely in both directions. The visible spectrum ranges from roughly 400 nm to roughly 700 nm. Light waves with wavelengths slightly shorter than this fall in the ultraviolet region. Toward shorter and shorter wavelengths, light falls in the X-ray and gamma-ray regions of the spectrum. Gamma rays have the shortest wavelengths.

In the opposite direction, as we move to wavelengths slightly longer than red, we reach the infrared region of the spectrum. Infrared radiation corresponds to heat, just like your body gives off. Regions of longer wavelength correspond to microwaves and radio waves. While it is

> 1 nm = 10^{-9} m = one billionth of a meter.
> 1 Å = 10^{-10} m = one ten-billionth of a meter.
>
> There are 10 Angstroms in 1 nanometer.

common to think that radio waves are sound waves, they are in fact electromagnetic waves. The electronic device we refer to as a "radio" is actually receiving these waves of radio "light" transmitted by a radio station, and converting them into sound for our ears to hear.

Another characteristic of a wave is its frequency (ν "nu"). The frequency can be thought of as the number of peaks that pass by a given point in one second's time. Picture a train with a long chain of identical cars attached to it. The "wavelength" of this train is the distance from one car hitch to the next. As the train passes by, you can count the number of car hitches that pass you in one second—this would be its frequency. While even a fast train will have a frequency, on this scale, of fewer than 5 cars per second, a visible light wave typically has frequencies of around 10^{14} cycles per second or more—that's 100 trillion peaks passing by every second! The unit used for frequency is called the hertz, and is abbreviated Hz (1 Hz = 1 cycle per second), as shown in **Figure 6.2**.

If you think about it, and see illustrated in **Figure 6.2**, these two characteristics are inversely related. As a light wave's wavelength increases, its frequency decreases. Since the peaks are more spread out, it takes longer for each to pass you by, meaning that in one second's time, fewer peaks pass by. In the case of the train, if the engineer wanted to maintain the same frequency in the event that his cars got longer, he could simply speed up. However, you have already learned that light travels at a constant speed: about 300,000 km/s. *All forms of electromagnetic radiation travel at the speed of light, regardless of their wavelength or frequency.* Even though an X-ray wave is oscillating furiously at an extremely high frequency, its speed through space is the same as a longer-wavelength radio wave, which has a much lower frequency.

Finally, a light wave's energy is directly related to its frequency. Despite the fact that all forms of electromagnetic radiation travel at the speed of light, X-rays and gamma rays have a much higher energy than microwaves and radio waves because they have higher frequencies.

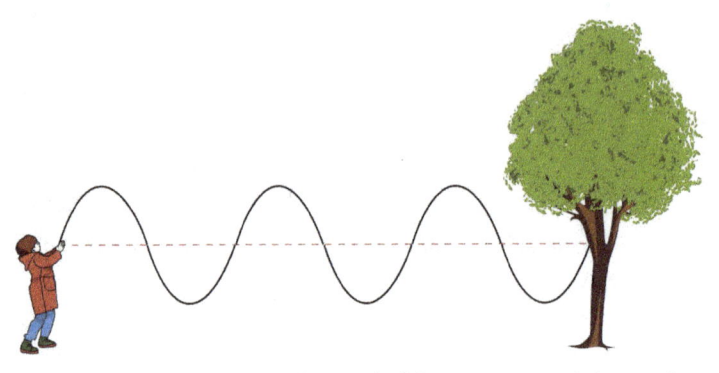

Figure 6.4. In this analogy, as the child shakes her end of the rope up and down, she creates a standing wave. The faster she shakes the rope, the more energetic the wave and the higher its frequency. If she shakes it more slowly, a lower energy, lower frequency wave is created. The energy in a light wave also directly corresponds to its frequency.

Copyright © 2009 by CK-12 Foundation / Wikimedia Commons / CC BY-SA 3.0

In **Figure 6.4**, a child is holding the end of a rope, with the other end tied to a tree. If the child shakes her end of the rope up and down smoothly, she can produce a wave in the rope with a specific wavelength and frequency. If she wants to create a wave with a higher frequency (shorter λ), she needs to shake the rope faster. It takes her more energy to shake the rope faster, and the wave she creates has more energy as a result. Conversely, if she shakes the rope more slowly, she exerts less energy and produces a lower-frequency (longer λ) wave. While this is only an analogy, it may be useful as you try to remember the correlations between wavelength, frequency, and energy.

6.2 The matter of little things

As Albert Einstein was working out the details behind his theory of relativity, another revolution was under development. Experiments had begun around the turn of the 20th century to study the structure of the most fundamental element of matter—the atom.

Preliminary work had revealed two components to the atom, one positively charged and one negatively charged. Under normal circumstances these two components were bound to each other electromagnetically, but under certain conditions the negative component could be removed. This was the foundation for the study of electricity and currents. In some cases, the atom itself was unstable and seemed to spontaneously emit one of these charged components. This phenomenon, known as radioactivity, was studied with special interest.

In 1897, a British physicist named J. J. Thomson discovered the true nature of this negative component, demonstrating that they were negatively charged particles that came to be known as electrons. Knowing that atoms were typically neutral, he figured the rest of the atom must be composed of the positively charged component. Considering these two things, he developed what became known as the "plum pudding" model (shown in **Figure 6.5**), where negatively charged electrons were embedded and distributed throughout a positively charged substance.

In 1911, Ernest Rutherford reported on the results of an experiment where helium nuclei were shot at a thin sheet of gold foil. He found that while most of the nuclei passed through the film and were detected on the other side (indicating that atoms were mostly empty space and not a "pudding" of positive charge), occasionally one of the nuclei was deflected backward off the foil at a very strong angle. By measuring the deflection angles and counting the frequency with which such deflections occurred, Rutherford determined that the obstacle must be quite tiny compared to the

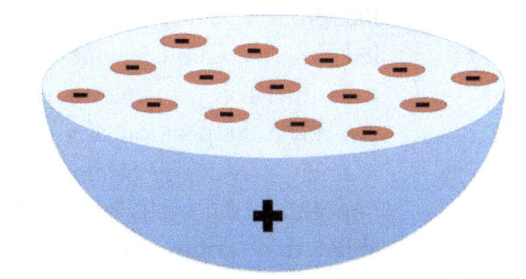

Figure 6.5. The "plum pudding" model of the atom, where negatively charged electrons (pink) were thought to be embedded within a positively charged substance. This model was later shown to be inaccurate.

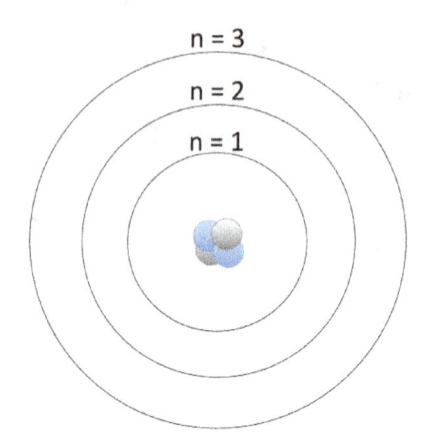

Figure 6.6. In the Bohr model, electron energy levels are drawn to look like planet orbits, where each increasing value of n corresponds to a higher amount of energy possessed by the electron. The $n = 1$ level is the ground state, the lowest energy the electron can have and the level the electron always tends toward. It is worth noting that this does NOT correspond to a physical diagram of the electron location(s); it only describes electron energies.

atom itself, yet massive, and positively charged. This positively charged nucleus must be surrounded by electrons located at relatively great distances, which orbit about the nucleus.

Following quickly on that discovery, Neils Bohr proposed that electrons must have certain specific energy values. The laws of electromagnetism predicted that if an electron is in an orbit, which undergoes acceleration since its direction of motion is constantly changing, then it ought to be constantly radiating away energy. This radiation should cause the electron to spiral inward, which clearly was not happening. Bohr's model explained that electrons could exist only in quantized energy levels, and that electrons could transition among these energy levels by absorbing or emitting energy. According to Bohr, there existed a lowest level ($n = 1$), called the ground state, where electrons would tend to go and exist in a stable state. Higher energy levels are referred to as excited states.

A simplified, *and physically inaccurate*, description of this (shown in **Figure 6.6**) resembles a solar system of planets orbiting around the Sun. In a Bohr model diagram like this, each energy level is drawn at increasing distances from the nucleus to indicate that the level has a higher energy value, with the lowest allowed energy value being closest to the nucleus.

These realizations about the nature of atoms were eye-opening to physicists at the time, but another surprise was still in store. As the 1920s progressed, a French physicist named Louis de Broglie made an astounding proposal. Not only does light behave both as a particle and as a wave, but matter does as well! In his 1924 doctoral thesis he was able to show that while electrons can be conveniently considered particles, they also exhibit wave properties—just like photons. The *particle–wave duality* of matter astonished the physics world and ultimately led to an incredible insight by Werner Heisenberg now known as the Heisenberg Uncertainty Principle. This principle states that because particles also behave as waves, you can never truly pin down the position of a particle without losing knowledge about its other characteristics (namely, its velocity or momentum). This is not due to technological limitations; it is a limit imposed by nature. Likewise, if you try to precisely measure a particle's motion, you lose the ability to determine its location.

Consider how we are able to make observations of anything. Imagine you are in a dark room and somebody tells you there is an object on the floor near you. You are given the task of determining the size and shape of this object and have at your disposal a number of rods of equal length but differing thickness. To improve your odds of finding the object at all, you ask for the thickest rod, a baseball bat, and then start poking around. Eventually the bat makes contact with the object and you can tell it is fairly flat, longer than it is wide, and one surface feels tacky. However, because the bat is so thick, you are unable to discern the finer details of the object.

> **Resolving Power**
> The ability for a measuring device to discern small details or distinguish between two objects.

In order to measure the object's finer details, you ask for the thinnest rod—a sharpened pencil. Using the sharpened point, you investigate the object and discover that it is rectangular and that the tacky side is actually composed of small squares of a material that feels rubbery. Comparing this to what you have experienced in your life, you are able to figure out that the object is a television remote control. In order to determine this, you needed a finer measuring device.

In physics and astronomy, the "measuring device" is usually light. Whether we are observing light emitted by distant objects or collecting light reflected off nearby objects, photons we receive from an object tell us about the object. If the object is small enough, we may need to use photons in a laser to probe an object's surface. As you have just learned, light can come in a variety of wavelengths. If the wavelength is too large, it acts like our baseball bat and prevents us from *resolving* much detail. In order to resolve small features, we must use light with a sufficiently small wavelength. This means that in order to detect a small particle, we would need to use short-wavelength light, such as X-rays.

However, you also just learned that short-wavelength light has a lot of energy. If we shoot a beam of X-rays at a particle to measure its position, the energy imparted on that particle by the X-ray photons has a significant impact on the motion (velocity) of the particle. On the other hand, if we try to reduce the impact on the motion by using low-energy light like radio waves, our ability to resolve the particle (its position) is greatly reduced as well.

One result of the Heisenberg Uncertainty Principle is that we cannot really say "where" an electron is in its orbit around an atomic nucleus. Erwin Schrödinger expressed this idea a year before Heisenberg by saying that particles have a "wave function"—an equation that describes the position of a particle as being more of a probability. A particle has a location of highest probability at any given moment and locations of lower probability. The principles of Einstein, Schrödinger, Heisenberg, and de Broglie opened up the field of quantum mechanics.

Within the atom, electrons do not inhabit definitive orbits as shown in the Bohr model. Instead, quantum mechanics says that the distribution of electrons around the nucleus is more like a "cloud" of probability (**Figure 6.7**). Within this cloud, an electron has a location of highest likelihood but can exist in any

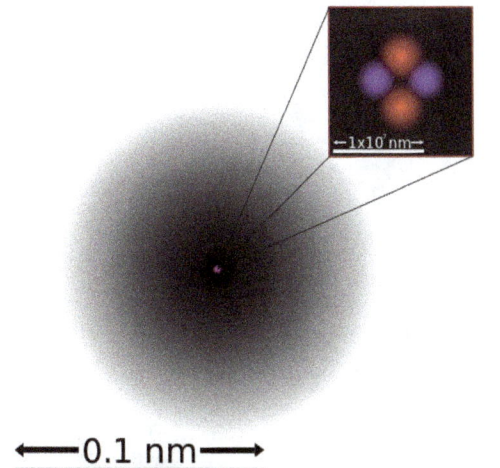

Figure 6.7. The electron cloud model of the atom, illustrating the concept that particles do not have a definitive location but exist in space with varying degrees of probability.

Copyright © 2008 by User:Furmanj / Wikimedia Commons / CC BY-SA 3.0

other place as well with varying degrees of probability. The Uncertainty Principle even says that the energy of an electron is not necessarily pinned down to a specific value but is also a function of probability, with a small degree of fluctuation (uncertainty) around the value given in the Bohr model.

With the advent of quantum mechanics, the scientific world was shaken a bit. For centuries it had been assumed that everything that was studied had definitive answers. In quantum mechanics, it seems that the most fundamental aspects of nature are instead more driven by statistical likelihood. In quantum mechanics there is no zero probability, only a very, very low degree of likelihood. This means that anything that is not mathematically forbidden can potentially, if given enough time, occur!

Even the most brilliant minds of the time, Einstein among them, looked upon quantum mechanics with skepticism. Einstein went so far as to say "I, at any rate, am convinced that [God] does not throw dice." Though quantum mechanics was met with resistance by many, its ability to accurately predict and describe results from particle physics experiments soon led

Figure 6.8. The Periodic Table of Elements.

Copyright © 2008 by Le Van Han Cédric / Wikimedia Commons / CC BY-SA 3.0

to widespread acceptance. Today, modern technologies like computer processors and medical imaging devices rely on the principles of quantum mechanics to operate. Like it or not, our world seems to be a very "uncertain" place!

Scientists now know there are many unique atoms in existence. These atoms have been arranged according to their electron configurations to form the periodic table of the elements, shown in **Figure 6.8**.

Each element in the periodic table differs from the others based on the number of positively charged protons it contains in the nucleus. This number corresponds to its atomic number. Other than hydrogen, whose nucleus consists of one lone proton, the nuclei of elements also contain particles called neutrons, which have no charge and are nearly the same mass as a proton. Since like charges repel one another, a nucleus made up of only protons would not be stable. The presence of neutrons in the nucleus helps keep the nucleus bound together by increasing the strength of the *strong nuclear force*. The strong nuclear force is one of the fundamental forces of nature that exerts itself over atomic and subatomic scales, binding protons and neutrons together in an atomic nucleus.

Some elements can exist with neutron numbers that are different from "normal." For example, a typical hydrogen atom has a nucleus of one lone proton. However, under certain conditions it can obtain a neutron, giving it an atomic mass of 2 instead of 1. It is still hydrogen, though, because its number of protons has not changed. This form of hydrogen is abundant enough in the Universe that it has its own name: Deuterium. Remember, though, deuterium is just another form of hydrogen. These two different versions of hydrogen are called *isotopes*, shown in **Figure 6.9**.

Sometimes an atomic nucleus can obtain too many neutrons. Such is the case in the aftermath of an exploding star, called a supernova. Atomic nuclei become bombarded by free neutrons to the point that many of them stick. There comes a point when the nucleus has too many neutrons and it becomes unstable. When a nucleus becomes unstable, the *weak nuclear force* can either cause one or more of the neutrons to be ejected or cause the nucleus to emit an electron, transforming one of the neutrons into a proton.

Under normal conditions, an atom is electrically neutral. The number of electrons it has is equal to the number of protons. The kinetic energy of the atoms—related to

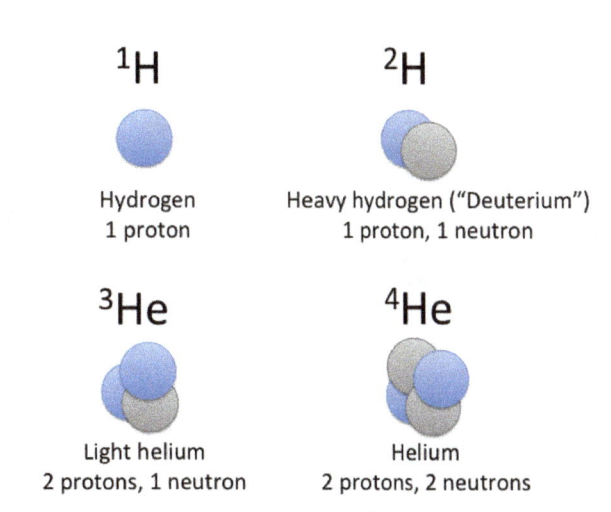

^{1}H
Hydrogen
1 proton

^{2}H
Heavy hydrogen ("Deuterium")
1 proton, 1 neutron

^{3}He
Light helium
2 protons, 1 neutron

^{4}He
Helium
2 protons, 2 neutrons

Figure 6.9. As long as the number of protons remains the same, the element remains the same. Adding more protons changes the element type. Adding more neutrons changes the isotope. Some isotopes, like ^{2}H (hydrogen-2), are stable. Others, like ^{13}C (carbon-13), are unstable and can be radioactive.

their motion—drops when atoms are relatively cool. The atoms within a substance slow down in their movement to the point where they can bind together to form *molecules*. Most atoms around you are locked into molecules of one form or another.

As a collection of atoms and molecules is cooled, energy is being removed from the system of particles. In theory, if we could remove all the energy from the system, cooling it until it could not be cooled any further, its temperature would reach a point called *absolute zero*. This zero point is defined on a temperature scale called the Kelvin scale, where absolute zero corresponds to 0 K. On the Fahrenheit scale, this is equal to –459.67° F (see **Figure 6.10**). Astronomers use the Kelvin scale because it allows them to specify the temperatures of objects relative to some objective, well-defined zero point, rather than using a rather arbitrary zero point like the Fahrenheit and Celsius scales use. Since temperature at its root is directly related to energy, then, it makes sense to define the zero point of a temperature scale in a way that corresponds to zero energy.

At its coolest, a collection of molecules locks itself into a solid; it freezes. We usually think of frozen things as being extremely cold, but our concept of "cold" is subjective. Some materials have freezing points at temperatures we would consider to be quite high—most metals, for example, freeze at temperatures well above the boiling point of water.

Gradually turning up the temperature, the solid melts and the molecules are able to move a bit more freely. However, the molecules are still somewhat attracted to one another, producing a liquid that flows but remains fairly cohesive. Heating up this liquid further, there comes a point when the liquid boils, turning our molecules into a gas. Solids, liquids, and gases are the typical phases of matter we experience in everyday life.

In astronomy, as you've already seen, circumstances rarely resemble our everyday life. To better understand the phases of matter in many astronomical scenarios, let's keep turning up the temperature on our molecular gas.

As the gas heats up, the molecules move faster and faster. With the increasing speeds, the molecules collide with each other more and more. There comes a point when the molecules in the gas collide with such force that they break apart, a process called *dissociation*. Our gas of molecules turns into a gas of atoms.

These atoms are still colliding with each other, and as we continue heating up the gas, these collisions begin to strip electrons off the atoms, *ionizing* the gas. Now this gas is composed of atoms that, lacking a complete set of negative electrons, now have a positive charge. These are called ions. Ions

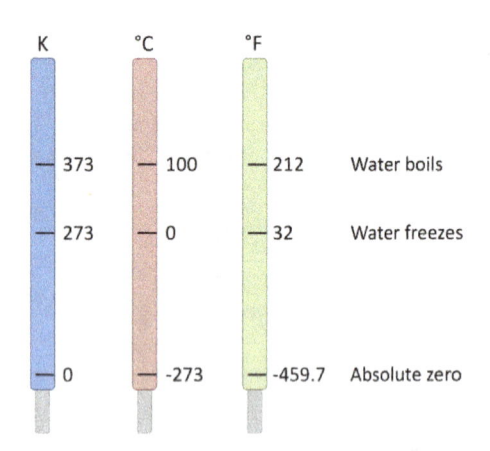

K	°C	°F	
373	100	212	Water boils
273	0	32	Water freezes
0	-273	-459.7	Absolute zero

Figure 6.10. The Kelvin temperature scale is favored in the physical sciences because it provides a natural zero point, not an arbitrary one. The Celsius scale is defined according to the characteristics of water.

and free electrons fill the gas. This state of matter is called a *plasma* and occurs in a hydrogen gas at temperatures around 10,000 K and higher.

As we continue to turn up the temperature, the ionized atoms continue losing electrons. By the time the temperature is in the millions of Kelvins, the last few stubborn electrons are stripped off and the atoms become fully ionized. The gas is now composed of free electrons and bare atomic nuclei.

Aside from small, cold objects like planets, comets, and asteroids, matter actually rarely exists as a solid. To create a solid generally requires both high densities (a lot of atoms in a small volume) and low temperatures (slow particle motions). In most astronomical scenarios we see high temperatures associated with high densities, and low temperatures with low densities. Some regions of space can have low densities and high temperatures as well. When densities are low, particles cannot come together to form solid objects, and when temperatures are high the gas typically ionizes. This is why the majority of atomic matter in the Universe exists in the form of stars (high temperatures and high densities) or gas clouds called *nebulae* (low densities and either high or low temperatures, depending on certain circumstances).

6.3 Interaction between light and matter

The whole field of astronomy can be understood in terms of the ways that light and matter interact. This makes the next two sections of this chapter a fairly important one, and you may want to review them from time to time as you read subsequent chapters.

There are four primary ways that light and matter can interact:

1. Matter can *emit* light
2. Matter can *absorb* light
3. Matter can *transmit* light
4. Matter can *reflect/scatter* light

In **Figure 6.11**, you can see how these appear in everyday life. Which of the above processes can you identify?

The way(s) that matter interacts with light is determined by the light's energy and the matter's composition—that is, what types of atoms are present and what electron energy levels are available.

As you have already learned, an electron within an atom can have only specific amounts of energy. These energy levels, shown in **Figure 6.12**, act like rungs on a ladder—in the same way that a person cannot stand between ladder rungs, an electron cannot have an energy value in between the allowed levels. **Figure 6.12** is identical to **Figure 6.6**, by the way. They are simply two slightly different ways of depicting the same general concept.

Figure 6.11. The above image illustrates the many ways that light and matter interact. Which of the four processes listed here can you see in this image?

Copyright © 2007 by User:rick / Flickr / CC BY 2.0

For an electron to jump up from the ground state to an excited state, it must absorb energy in the process. There are several ways this can happen, but two of the most common ways are through collisions and light absorption. If atoms collide, some of the kinetic energy

Figure 6.12. The electron energy level diagram for the hydrogen atom. Hydrogen generally has just one electron, which can exist in any of the given energy levels but will usually seek the ground state. By convention, the energy of each level is indicated with a negative sign and in units called *electron-Volts (eV)*. The point where the electron escapes the atom, causing the atom to become ionized, is often referred to as $n = \infty$ and corresponds to 0 eV on this scale. Many levels exist between $n = 4$ and $n = \infty$, but have been omitted for clarity.

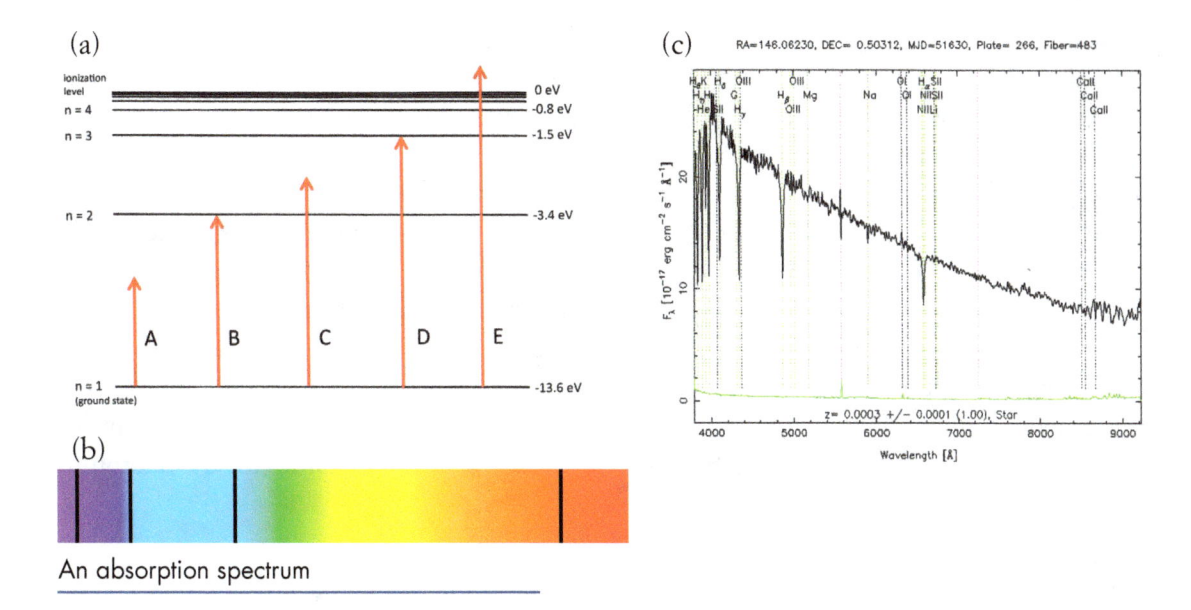

Figure 6.13. The process of light absorption, where only photons of specific energies can be absorbed. (a) The transitions labeled "B" and "D" are allowed, meaning light with sufficient energy to span the "energy gap" can be absorbed. However, "A" and "C" are *not* allowed since they do not result in an electron ending up at an existing energy level. Very high-energy photons can be absorbed to the point where the electron is ejected ("E"), ionizing the atom. (b) When certain wavelengths (energies) of light are absorbed, a black line is produced in a spectrum. This is called an absorption line. (c) When an absorption spectrum is plotted on a graph, these absorption lines correspond to points that dip down below the overall continuum.

Public Domain; Copyright © 2009 by SSDS Collaboration (www.sdss3.org). Reprinted with permission.

in the impact can be transferred to the electrons, boosting them up to higher energy levels. Alternatively, photons can be absorbed by the electrons within the atoms, providing a similar boost.

However, not just any photon can be absorbed. **Figure 6.13a** illustrates the process of absorption, where an electron starts off in a lower energy level and jumps to a higher level if the photon has the right amount of energy to allow it to cross the "energy gap." If the photon's energy isn't just right, the photon isn't absorbed.

For example, the arrow labeled "B" in **Figure 6.13a** corresponds to an electron that jumps from the ground state, with an energy of −13.6 eV, to the $n = 2$ first excited state, where its energy is −3.4 eV. This means that the absorbed photon must have an energy of 13.6 − 3.4 = 10.2 eV.

On the other hand, if the incoming photon has an energy of only 8.0 eV (indicated by the arrow labeled "A") then the photon will not be absorbed because that transition does not place the electron at an existing energy level. If the photon has an energy of 12.1 eV, the electron in the ground state can absorb it and jump up to the $n = 3$ second excited state (indicated by the arrow labeled "D"). Other transitions are allowed, as you can see, and if the photon has an energy of 13.6 eV or higher (indicated by the arrow labeled "E"), then it can be absorbed by that electron to the point where the atom becomes ionized.

Once an electron has been excited, it will not remain in the excited state for long. Fairly quickly, the electron will want to return to the ground state—but to do so, it will need to get rid of the energy it gained. To make this possible, a photon is emitted that has the energy corresponding to the transition the electron makes.

As you can see in **Figure 6.14a**, if the electron is in the $n = 2$ first excited state, it will emit a photon of 10.2 eV to return to the $n = 1$ ground state. It cannot emit a 9.0 eV photon because this will not allow the electron to return to an existing energy level. An electron in the $n = 3$ second excited state will emit a photon with 12.1 eV to return to the ground state. Alternatively, it could emit a 1.9 eV photon to go from $n = 3$ to $n = 2$, and then emit a 10.2 eV photon to

An emission spectrum

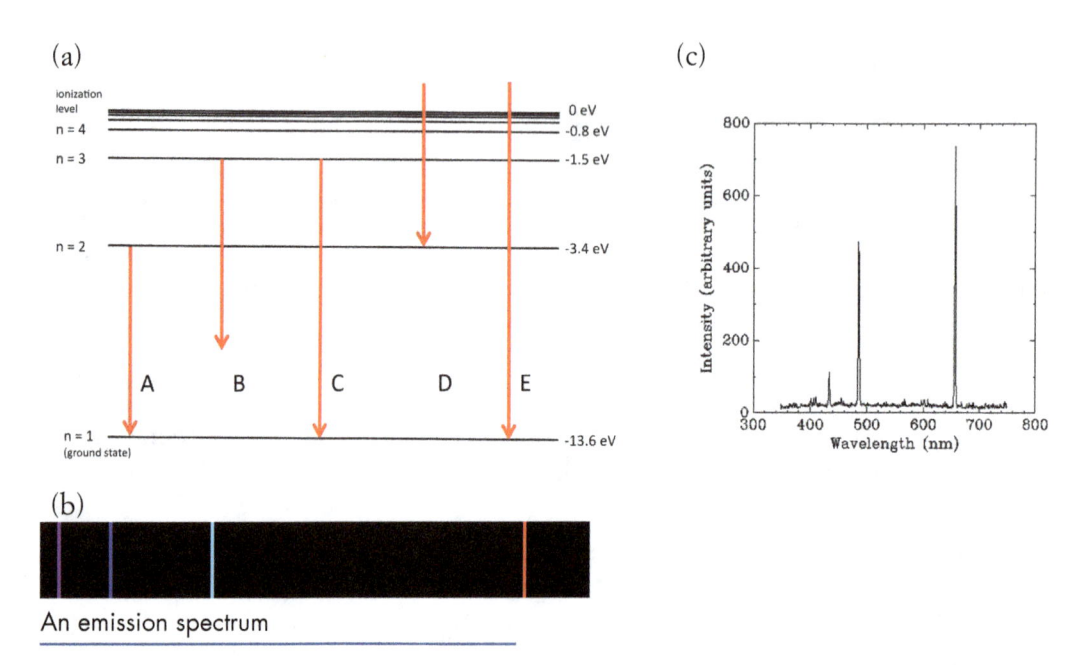

Figure 6.14. The process of light emission, where only photons of specific energies can be emitted. (a) The transitions labeled "A" and "D" are allowed, meaning light with energy corresponding to the "energy gap" will be emitted. However, "B" is not allowed since it does not result in an electron ending up at an existing energy level. Free electrons can recombine with ionized atoms and emit enough energy to land at any existing energy level (labeled "D" and "E"). In the case of "D," the electron will first emit a 3.4 eV photon to reach the $n = 1$ state, and then possibly emit a 10.2 eV photon to drop down to the ground state afterward. (b) When certain wavelengths (energies) of light are emitted, a colored line is produced in an otherwise colorless spectrum. This is called an emission line. (c) When an emission spectrum is plotted on a graph, these emission lines correspond to points that rise up above the background or continuum.

Copyright © 2007 by User:Profeta / Wikimedia Commons / CC BY-SA 3.0

go from $n = 2$ to $n = 1$. Whether the former transition occurs or the latter is determined by statistical likelihood.

Comparing **Figure 6.13b** to **Figure 6.14b**, you may notice that the dark absorption lines in **Figure 6.13b** match the colored lines in **Figure 6.14b**. This is going to be true when comparing *absorption and emission spectra* from substances with the same composition. The spectra in **Figure 6.13b** and **Figure 6.14b** are both from hydrogen gas, one where the hydrogen is absorbing light and the other where hydrogen is then re-emitting the absorbed light. The transitions available for light absorption are the same transitions available for light emission. Most importantly, this pattern of lines is unique to hydrogen. This means that if we were to obtain a spectrum from an object and see this pattern of lines in it, we would know that hydrogen was present in the object.

Spectra therefore allow us to determine chemical composition. Biologists and chemists use the technique of *spectroscopy* to study chemical reactions and structure, and astronomers use spectroscopy to study stars, galaxies, and even the space between stars and galaxies. The development of spectroscopy has been one of the most important and valuable tools astronomers have for studying and understanding phenomena in the Universe.

As illustrated in **Figure 6.15**, light from a distant object is broken up by wavelength using an optical element like a prism or diffraction grating. This spectrum is then recorded using an imaging chip called a *charge-coupled device (CCD)*, which is essentially a science-grade version of the imaging chip in your digital camera or camera phone. The image of the spectrum can then be measured to produce a graphical version similar to those shown in **Figures 6.13c** and **6.14c**.

Modern spectroscopy has produced spectra with extremely high resolution, often at less than one Angstrom per pixel. By obtaining such fine detail, astronomers can extract a large amount of information about the target object.

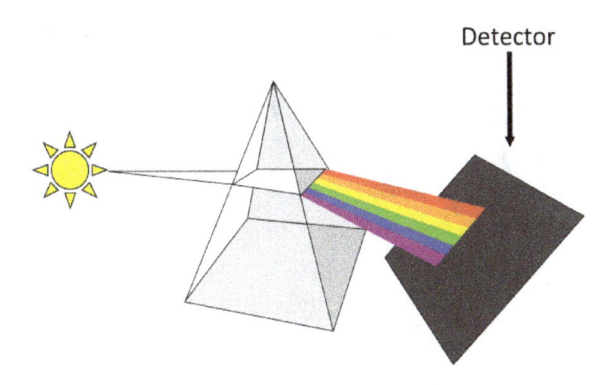

Figure 6.15. In spectroscopy, light from a distant source is broken up into a spectrum using an optical element like a prism. The spectrum is then recorded on CCD for measurement.

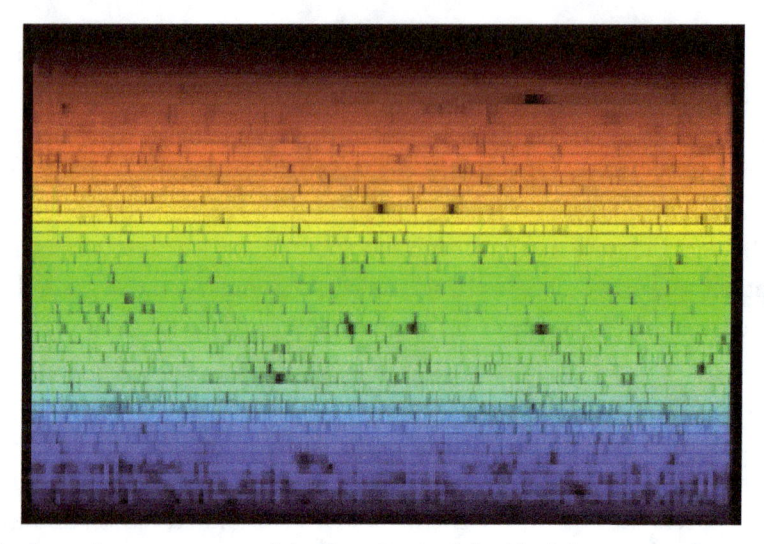

Figure 6.16. A high-resolution spectrum of the Sun. Each of the black features in the spectrum is an absorption line produced by different atoms and molecules in the outer atmosphere of the Sun.

N.A.Sharp, NOAO, NSO, Kitt Peak FTS, AURA, and NSF. Copyright © 1984 by National Optical Astronomy Observatory (NOAO). Reprinted with permission.

Figure 6.16 shows a high-resolution spectrum of the Sun. Each of the absorption lines in the spectrum corresponds to a unique fingerprint from atoms and molecules present in the Sun's outer atmosphere. The presence or absence of various lines, along with their relative strengths, also tells astronomers about environmental conditions like pressure, temperature, and magnetic field strength. Spectroscopy reveals an abundance of information that imaging alone could not.

The background of **Figure 6.16** and **Figure 6.13b** illustrate the third type of spectrum, called a continuous spectrum. To understand how a continuous spectrum is formed, imagine taking a substance or process that is producing light, such as nuclear fusion reactions or the stimulated emission of radiation that occurs in light bulb filaments or lasers. These processes produce photons of only specific energies, which would normally appear as an emission spectrum.

However, in certain conditions (like the Sun's core) the density of particles is extremely high. As soon as photons are produced, they collide with and scatter off nearby particles. These collisions change the energies of the photons in a largely random way that is related to the average energy of the particles—the temperature. As the photons eventually escape from this hot, dense object, their energies have been so randomized that instead of producing a number of distinct emission lines, the total light emitted constitutes a broad, continuous distribution of energies. In the visible spectrum, this appears as a rainbow. On a graphical representation of a spectrum, as in **Figure 6.17**, this looks like a smooth curve often referred to as a continuum.

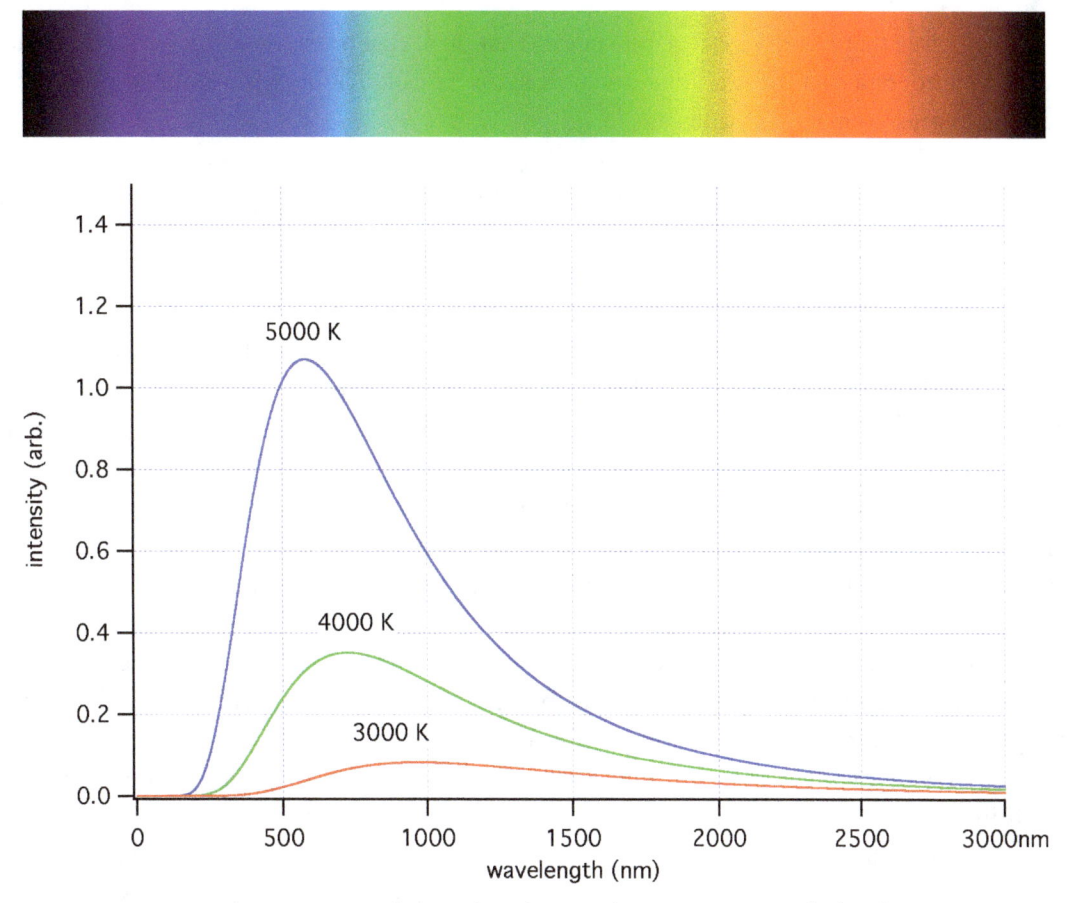

Figure 6.17. A continuous spectrum has no breaks in it, but is just a smooth distribution of light at all wavelengths.

Because a *continuous spectrum* is related to the kinetic energy of particles in a light-emitting object, the characteristics of the spectrum are related to the temperature of the object. When an object is hotter, such as a very hot star, the particles in its interior have much higher energies. This means the particles are moving faster, so the energy transferred to photons during collisions is, on average, much higher. This means that the average energy of the photons that escape from this hot star is high as well. The result of this is that the peak of the continuum is at a higher energy—meaning a shorter wavelength, shown in **Figure 6.18** by the curve marked "12,000 K."

In cooler stars, on the other hand, the particles are moving a bit more slowly. This means that the average energy of light escaping from the star is lower, producing a spectrum that peaks at longer wavelengths. This is reflected in the curve labeled "3,000 K" in **Figure 6.18**.

The principle that hotter objects emit spectra that peak at shorter wavelengths, while cooler objects emit spectra that peak at longer wavelengths, is called *Wien's* (pronounced "veen's") *Law*. It is expressed mathematically as

$$\lambda_{max} T = 2.9 \cdot 10^6 \; nm \cdot K$$

> **Wien's Law**
> Hotter objects emit light with a higher average energy, meaning their spectra peak at bluer wavelengths.

This brings up an important side note: our eyes perceive stars to have different colors due to their temperatures. However, what matters in this case is how much blue light is emitted compared to red light within the narrow visible light window illustrated in **Figure 6.18**. If you look at the 12,000 K star, it clearly emits more blue light than red light. Thus, it would appear blue to our eyes. The 3,000 K star emits more red light than blue light, so it would appear red to our eyes.

On the other hand, the 6,000 K star emits a spectrum that peaks near the middle of the visible window. Our Sun, which is a bit cooler still, has a spectrum that peaks in the green/yellow region. Although the Sun is often represented as yellow (an effect that results from our atmosphere scattering out the blue light before it reaches our eyes), if you could go out into space you would see that the Sun actually appears white. Why does it appear white, and not green? It's because our eyes are sensitive to all of these colors in a fairly comparable way at high intensity. This means that our eyes don't perceive there to be much of a difference between the amount of blue, green, yellow, and orange light received from the Sun, and all of it is quite bright. Our eyes and our brain work together and add it all up, producing white. In a sense, then, "color" is subjective. While there are no green stars as we would perceive them, there are stars that emit light that peaks in the green part of the spectrum.

A continuous spectrum that is produced in such a way that its shape and peak are related to the temperature of the emitting object is also referred to as a *blackbody spectrum*. A blackbody is a theoretical object that perfectly absorbs all radiation that strikes it, with no reflection, and perfectly radiates light over all wavelengths in a way that is dependent on its temperature. The radiation it emits is therefore sometimes called *thermal radiation*. While no real object emits a perfect blackbody spectrum, certain objects like stars emit light in a way that closely resembles a blackbody spectrum.

If you look at **Figure 6.18** once again, you will see that these spectra don't intersect. The line indicating the spectrum from a cooler star always remains below the line indicating the spectrum from a hotter star. While the graph shows only a small portion of the full electromagnetic spectrum, if we could see the entire range plotted, you would still see this point to be true. This fact is referred to as the *Stefan-Boltzmann Law*.

> **Stefan-Boltzmann Law**
> Hotter objects emit more light per unit area at all wavelengths than do cooler objects.

Light and Matter | 103

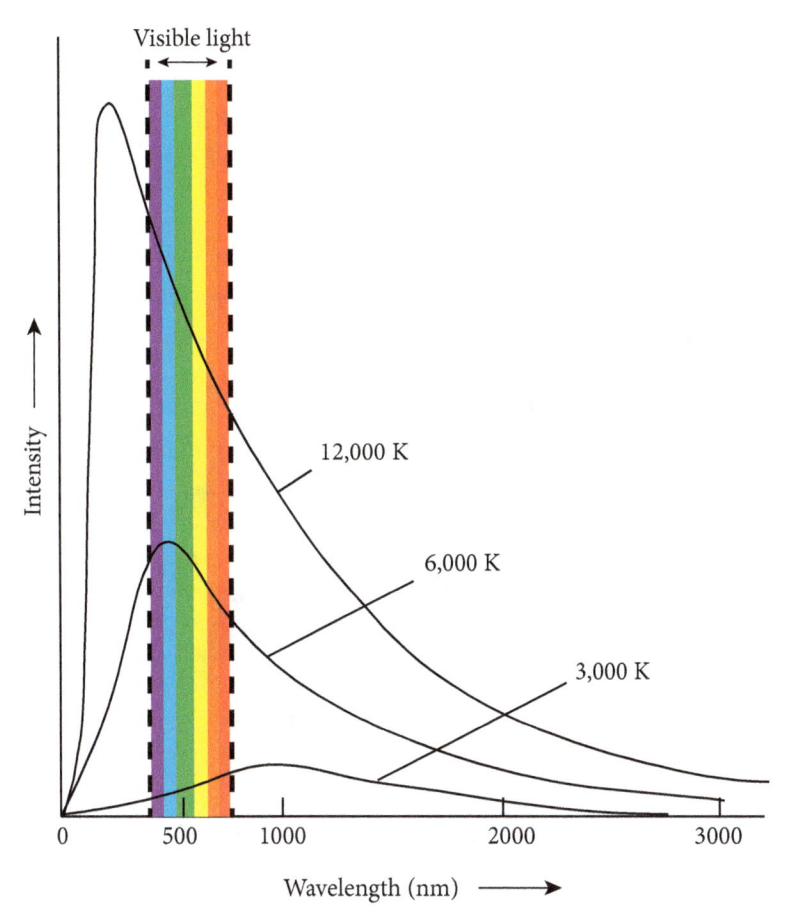

Figure 6.18. Three examples of continuous spectra produced by objects of different temperatures. The value along the vertical axis is actually "intensity per unit surface area." Notice how the peak of the spectrum shifts with temperature. This illustrates the concept of Wien's Law. How do you think the spectrum emitted by a human body would look on this diagram?

User:Drphysics / Wikimedia Commons / Public Domain

Tools of the Trade 6.1

Wien's Law

$$\lambda_{max} T = 2.9 \cdot 10^6 \; nm \cdot K$$

Looking at this equation, it might look intimidating at first. The two factors being multiplied on the left side of the equation are λ_{max}, which represents the wavelength of the peak of the continuous spectrum (in nm), and T, the temperature of the object (in K). Consider the units that would result from multiplying a number in nanometers by a number in Kelvins—the resulting product would have units of "nanometers times Kelvin," or "nanometer-Kelvins." This is why the number on the right side of the equation has these units.

This equation says that whenever we multiply the peak wavelength by the temperature, in the units of nanometers and Kelvins, we get the same number: 2.9×10^6, or 2,900,000.

Consider what combinations of numbers are needed to be multiplied together to equal 8. Here are the options:

$$8 \times 1 = 8$$
$$4 \times 2 = 8$$
$$2 \times 4 = 8$$
$$1 \times 8 = 8$$

Notice that in order for the product to remain the same, as the first number is decreased the second number must increase accordingly. The same is true with Wien's Law.

In order for the multiplication in Wien's Law to always produce the same number, as the temperature of the object increases, the peak of its emitted spectrum must shift to shorter (bluer) wavelengths:

λ_{max} (nm)		T (K)		(nm · K)
967	×	3,000	=	2,900,000
483	×	6,000	=	2,900,000
322	×	9,000	=	2,900,000
242	×	12,000	=	2,900,000

Conversely, as the temperature of the object decreases, the peak wavelength shifts to larger (redder) values.

Using this equation, and a little algebra, you can calculate the temperature of an object if you know the peak wavelength of its emitted spectrum, or vice versa.

The Stefan-Boltzmann Law states that each unit area of a hotter object's surface emits more light at every wavelength than that of a cooler object. This means that if you could mark off one square foot (or one square meter, or one square kilometer…) on the surface of the 12,000 K star shown in **Figure 6.18**, not only would that square foot emit more blue light than a square foot on the surface of the 3,000 K star but it would also emit more red light. It would also emit more X-ray light, and more infrared light, more radio light, etc. This also means that if we had two objects that were the same size, but different temperatures, the hotter object would not only appear bluer (Wien's Law) but it would also appear brighter (Stefan-Boltzmann Law). You can try this with a stovetop burner (**Figure 6.19**) or a toaster—watch how the metal heating elements get brighter as they get hotter. On some appliances you can even see the filaments go from a deep red to a bright orange.

While the understanding of light, matter, and the interaction between the two took several centuries to fully develop, astronomers made use of the developing knowledge every step of the way. Not only did it lead to greater insight into the physical Universe, but each step also inspired technological innovations that made data gathering more capable and more efficient. In the next chapter, we will look at the tools and systems astronomers have developed for collecting and organizing data, as well as how the principles you have learned in this chapter have allowed astronomers to understand each new observation.

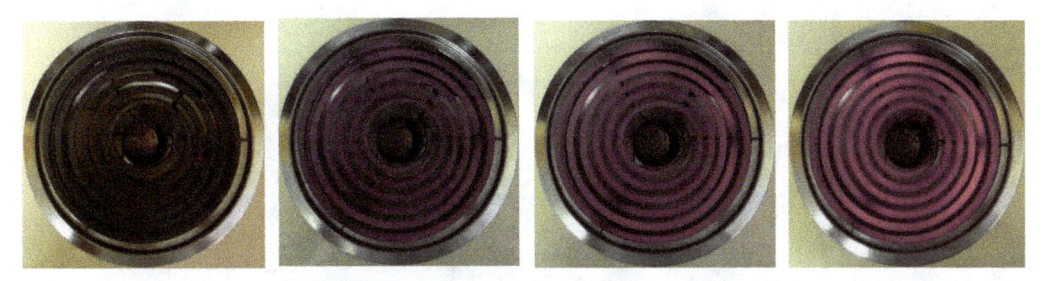

Figure 6.19. A stovetop burner illustrates the Stefan-Boltzmann Law and Wien's Law. As it heats up, the coil changes from dim to bright and goes from deep red to bright red. Some coils even get hot enough to turn orange.

CHAPTER SEVEN

The star cluster NGC 3766. Called an "open cluster," this group of stars is thought to be quite young—perhaps having formed less than 20 million years ago.

Copyright © 2013 by European Southern Observatory / CC BY 3.0

Measuring the Stars

Learning Objectives

In this chapter you will learn:

- » The primary functions of a telescope
- » To identify the three most common types of telescopes and draw light-ray diagrams for each
- » How astronomers define the magnitude system for apparent brightness
- » The ways that brightness and luminosity are related
- » The ways that astronomical spectra originate

Key Words

- Telescope
- Aperture
- Light-gathering power
- Focal length
- Objective lens
- Chromatic aberration
- Spherical aberration
- Interferometry
- Adaptive optics
- Luminosity
- Apparent magnitude
- Parsec

7.1 The role of light in astronomy

When you think about it, nearly everything in astronomy comes down to light*. Since we cannot fly to distant stars, we must study the light they emit or reflect to understand them. Planets orbiting around other stars obscure or reflect stellar light, and emit infrared light

---

* Astronomers can sometimes detect energetic matter particles that have traveled through space, as well.

of their own. Gaseous nebulae emit, scatter, and absorb light as well. When we observe a moving object, we see its light first coming from one location, and then another. Although using only light to understand the Universe may at first seem limiting, astronomers have developed many novel ways of using photons to "shed light" on celestial objects and phenomena.

Before there were methods for making and recording telescopic observations, astronomers produced great catalogs of star positions and maps of the night sky. By noting positions, they were able to deduce that not everything in the night sky remains still—objects such as the Sun, Moon, and planets all display motion against a seemingly static backdrop of stars. A nearby planet like Venus even demonstrates small but noticeable fluctuations in brightness, indicating that at some times it is relatively close to Earth, and at other times it is more distant.

While the accuracy of these catalogs was improved with each generation of stargazer, little more could be done until the development of instruments that could aid in observation. Telescopes, and the imaging devices that were subsequently developed, brought astronomy into an entirely new arena.

7.2 Telescopes

You have learned that while Galileo did not invent the telescope himself, he was the first to use one to view the night sky in a systematic way. The immediate impact that Galileo's observations had on the scientific community was indicative of the lasting value that the telescope would have in the field of astronomy.

The telescope itself has not remained a static instrument. Many changes and improvements have been made over several centuries, taking them from simple handheld devices to truly gargantuan constructions that dwarf many typical buildings. In this section, we will look at common telescope designs and see how they have been implemented.

At its simplest, a telescope is an instrument used to collect light emitted or reflected by an object. It does so by allowing light to enter through an opening called the *aperture*. In some types of telescopes the aperture is open, while in others it holds a lens. In either case, the larger the aperture, the more light the telescope can collect.

A telescope has several purposes, with light-gathering being its primary function. Hence, a telescope is often referred to by its aperture size, as this is a measure of its *light-gathering power*. For example, a telescope with an aperture diameter of one meter is often simply referred to as a "one-meter." A telescope with higher light-gathering power is capable of revealing fainter objects and structures than a smaller telescope.

A telescope is also designed to improve one's ability to resolve details of a distant object. This function, called its *resolving power*, increases with aperture size as well. This

> **Telescope**
> An instrument used to collect light emitted or reflected by an object.

means that large telescopes can see finer detail than smaller telescopes. An image of a fuzzy elongated shape seen through a small telescope could be resolved as two individual stars in a larger telescope.

This is not because of its increased light-gathering power, but is actually related to the size of the aperture compared to the wavelength of light being observed. The human eye can resolve details down to about one arcminute (1/60th of a degree). The ability to resolve fine detail decreases as the wavelength increases. This means that in theory, blue details should be easier to distinguish through a telescope eyepiece than red details. This is also why radio telescope dishes are so large (**Figure 7.1**)—the long wavelengths of radio waves require large aperture telescopes to resolve and distinguish features.

Finally, telescopes are often considered "magnifying instruments," allowing the user to see a distant object much more easily. This feature of a telescope, more accurately referred to as its *image scale*, is related to its *focal length*—the distance that the light travels from the first optical element it experiences (lens or mirror) to the point where it is focused. This is illustrated in **Figure 7.2**.

However, while the magnifying power of a telescope might seem incredibly important—and it is indeed useful—without the telescope's ability to resolve the detail being magnified by gathering a sufficient amount of light, the magnification is worthless. A magnified image with poor resolution simply becomes a blurry mess.

For eyepiece viewing, the magnification is also related to the focal length of the eyepiece being used. Mathematically, the relationship is given by

$$M = \frac{FL_{Telescope}}{FL_{eyepiece}}$$

Looking at this equation, if the telescope focal length in the numerator is increased then the overall fraction gets larger. This means that telescopes with long focal lengths are capable of providing high magnification. This is also why professional

Figure 7.1. A telescope's resolving power is related to its aperture and the wavelength of light to which it is attuned. A radio telescope, which observes very long wavelength light, must have a very large aperture to resolve details.

Geremia / Wikimedia Commons / Public Domain

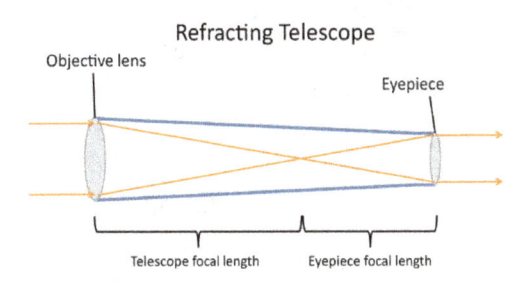

Figure 7.2. Basic optical design and light-ray diagram for a refracting telescope. The yellow lines represent the path that light rays take as they pass through the lenses. The objective lens collects the light and focuses it to a point over a distance referred to as the focal length. The light then travels out through the eyepiece, where the observer's eye focuses it onto the retina.

digital photographers use very large "zoom" lenses; the long focal lengths of these zoom lenses provide high magnification by improving the image scale.

On the other hand, if you increase the denominator—the focal length of the eyepiece—then the fraction gets smaller. When viewing objects through a telescope, an eyepiece with a longer focal length creates a lower-power wide field view. Zooming in therefore requires eyepieces with shorter focal lengths. A 20 mm eyepiece provides twice the magnification that a 40 mm eyepiece provides.

7.2.1 Refracting Telescopes

The first type of telescope invented was the *refracting telescope*, shown in **Figure 7.2**. Refracting telescopes operate by using curved, transparent glass lenses to refract, or bend, light to a focus. As light enters a telescope, it passes through the *objective lens*—the lens that sits at the aperture of the instrument.

If you follow the top light ray in **Figure 7.2** as it passes through the telescope and exits through the eyepiece, you'll notice that it ends up exiting at the bottom. This reflects the fact that telescopes produce images that are often inverted and/or mirrored. While this might cause some confusion viewing Earthly objects (an effect that is corrected for in binoculars), this is not an issue when viewing celestial objects because the concept of "up" and "down" lose their meaning in the vast expanses of space.

The objective lens is shaped to focus the light it collects to a point over a defined focal length. All optical elements, including lenses, mirrors, and even the human eye, have focal lengths. The human eye brings light to a focus on the retina, located in the back of the eye, where it is turned into an electrical signal interpreted by the brain as an image. The objective lens of a refracting telescope collects the light and directs it inward to a focus. It is then allowed to pass out of focus just enough for the eyepiece to collect it and redirect it outward, where the observer's eye can collect it again. In this way, telescopes are essentially instruments that can collect a large amount of light and deposit it into the human eye (or onto a camera) for processing.

All telescopes have advantages and disadvantages. It is up to the astronomer to decide what advantages he or she wants, and what disadvantages he or she is willing to accept or work around. With lenses used in refracting telescopes, different wavelengths (colors) of light are bent at slightly different angles. This means that the blue light collected from an object is not focused at the same exact point

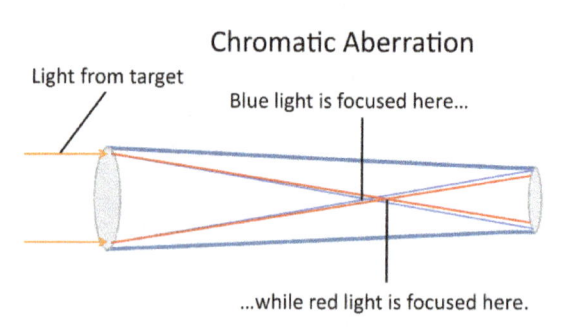

Figure 7.3. Telescopes that use lenses suffer from chromatic aberration, where different wavelengths of light are focused at slightly different points, causing color distortion in the image.

within a telescope as the red light. This is called *chromatic aberration*, shown in **Figure 7.3**. Because of chromatic aberration, an image of a star viewed through a simple refracting telescope can be slightly different colors, or can be colored differently at its extremity, depending on how the eyepiece is positioned. Usually this problem is addressed by creating eyepieces with multiple lenses shaped to make every color coincide as much as possible. If the effect can be corrected then refractors provide a high degree of contrast, allowing color and brightness variations in the image to be more easily seen.

7.2.2 Reflecting Telescopes

It was not long after the invention of the refracting telescope that another type of telescope was developed, which used mirrors instead of lenses to collect and focus light. These *reflecting telescopes* were initially designed to address the problem of chromatic aberration in refracting telescopes. Since the light does not pass through a lens, there is no chromatic aberration present in reflectors.

Isaac Newton is credited with designing the first reflector, implementing a primary mirror ground and polished into a parabolic shape that directed the light back up through the telescope tube. A flat secondary mirror near the aperture then directed the light out through a hole in the side of the telescope tube. This Newtonian design, shown in **Figure 7.4**, avoided chromatic aberration, but, by using a parabolic mirror, introduced a new form of aberration called *coma*. Coma results when light that enters the telescope "off-axis" (meaning not parallel to the axis of the telescope tube) is focused asymmetrically. Star images near the edge of the field of view become smeared out into a teardrop shape. This is typically avoided by using a relatively small aperture compared to the length of the telescope tube.

One advantage that a reflecting telescope design has over a refractor is that the primary mirror can be made much larger than a glass lens. In a refracting telescope, the lens is mounted around its perimeter and if the lens gets too large it can begin to sag, causing distortion. A mirror, on the other hand, needs to be polished on only one side and can be completely supported on the unpolished side. Thus, a mirror can be made to very large sizes without the same degree of distortion (although eventually a very large mirror also begins to sag if it is tipped enough).

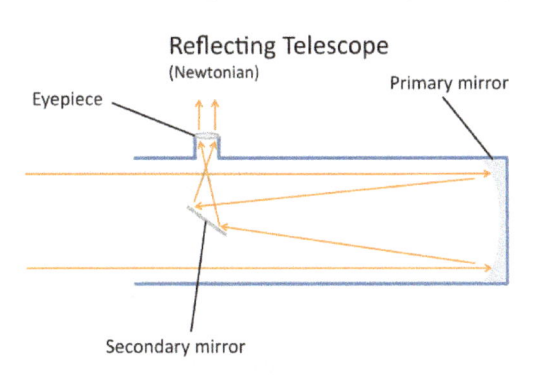

Figure 7.4. The Newtonian reflector design, where a curved primary mirror directs light to a flat secondary, which then redirects the light out through the eyepiece near the telescope aperture.

The second advantage a reflector has over a refractor is that it can be made more cheaply. A large glass lens needs to be of the utmost quality and clarity, since light passes through it. Because

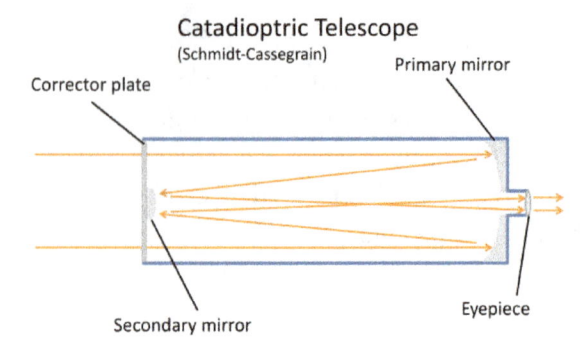

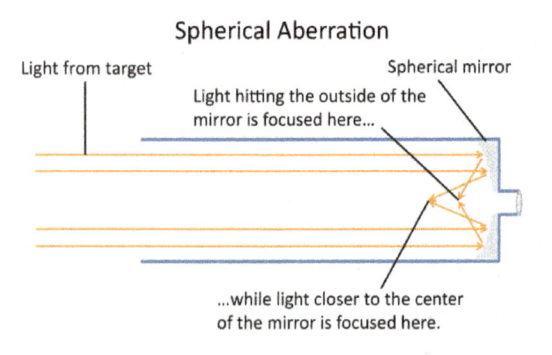

Figure 7.5. A catadioptric telescope (top) is a category of telescopes that use both mirrors and a lens to focus light. Without the corrector plate (lens), the telescope would suffer from spherical aberration (bottom).

mirrors only reflect light, they do not need the same degree of quality to provide optimal results.

The Newtonian design is just one of several ways of producing a reflecting telescope. One other popular design is called a Cassegrain reflector, where light is redirected through a hole in the primary mirror to the eyepiece. This design also appears in the next category of telescope.

7.2.3 Catadioptric Telescopes

A third category of telescope exists that utilizes both mirrors and a lens. As telescope-making improved, engineers realized that by creating a mirror with a spherical surface, rather than parabolic, coma could be eliminated. However, a spherical optical surface produces an aberration called *spherical aberration*, where light rays that strike the mirror near its edge are focused more closely to the mirror than those that strike near the center, shown in **Figure 7.5**. However, a correcting lens could be designed that deviates the light rays in such a way that this aberration could be eliminated.

In order to increase the magnifying power of the telescope, the focal length was increased by using the secondary mirror to direct the light back down the telescope tube, instead of out through a hole in the side, as in the Newtonian design. Compare **Figure 7.5** to **Figure 7.4** to see how the focal length has increased. The light can effectively travel down the full length of the telescope one more time before coming to a focus, all without having to make the telescope itself any longer. The light then passes through a hole drilled in the primary mirror, where the eyepiece is attached. This specific design is called a Schmidt-Cassegrain telescope and is a popular model for high-end amateur instruments. They provide quality viewing and high magnification in a compact form. However, the added corrector plate makes the telescope more expensive. Additionally, the corrector plate reflects some light, thus dimming the image slightly. Removing light from the beam that enters the telescope is not desirable for scientific study, so most professional science-grade telescopes are open-aperture Newtonian or Cassegrain models.

7.2.4 Modern Adaptations

As telescopes have gotten larger, astronomers have had to address some very practical limitations. At some point, a mirror is just too large to manufacture with sufficient precision, and if tipped too far it can sag under its own weight. Additionally, the increase in resolving power with aperture size reaches a maximum, where turbulence in the Earth's atmosphere prevents further gains by distorting the image.

To address these issues, astronomers and engineers introduced some clever innovations. The first step was to construct larger mirrors by assembling arrays of smaller mirror segments, as shown in **Figure 7.6** from the Keck Observatory on Mauna Kea, Hawaii. Crafting mirror segments that fit together allows astronomers to build mirrors ten meters in diameter and larger. To get a sense of what ten meters looks like, one stride for an average person is about a meter. Take ten strides across your room, or outside, to get a sense of how large these world-class telescopes can be. Plans are already in place to build even larger telescopes!

Astronomers were still not satisfied, however. A 10-meter telescope can surely provide a large amount of light-gathering power, and great resolving power, but they wanted to try improving the resolving power even more. To do this, sophisticated algorithms were introduced to combine light signals together from multiple telescopes. This process, called *interferometry*, allows the multiple telescopes to attain the same resolving power as that of one

Figure 7.6. The largest telescopes in the world utilize primary mirrors constructed out of multiple smaller segments.

Copyright © 2007 by User:SiOwl / Wikimedia Commons / CC BY 3.0

Figure 7.7. Light signals from multiple telescopes can be combined through interferometry to increase the resolution of the image, as is done at the Very Large Array (left) and the Very Large Telescope Interferometer (right). VLTI image courtesy of NRAO/AUI. VLT image courtesy of ESO/H.H.Heyer.

(left) Copyright © 2002 by NRAO/AUI / CC BY 3.0; (right) Copyright © 2007 by ESO/H.H.Heyer / CC BY 3.0

Figure 7.8. By using artificial stars created by lasers, astronomers can measure the atmosphere's effects on an image and subtract it out in a process called adaptive optics. Many of today's modern telescopes, like the two Keck telescopes and the Subaru telescope shown here, are outfitted with these systems to provide images with remarkably high resolution.

Copyright © 2006 Paul Hirst / Wikimedia Commons / CC BY-SA 2.5

telescope with an effective diameter equal to the separation distance of the individual telescopes. This technique is used in radio observatories like the Very Large Array (**Figure 7.7a**) and visible light observatories like the Very Large Telescope Interferometer (**Figure 7.7b**).

Despite these advances, the atmosphere is still the primary cause of a limit in resolving power. Atmospheric turbulence causes distortion in an image as the light passes through the air. Think back to what you learned about lenses. As air near the Earth's surface absorbs heat, its density decreases and it rises due to buoyancy. The difference in density makes it act like a lens, causing light rays to be refracted as they pass through it. Have you ever noticed how the pavement off in the distance seems to shimmer while you drive down a long road on a hot day? This is exactly the same effect. That shimmering air causes images to become distorted as it rises through the atmosphere.

While one solution to this problem is to launch telescopes into space where they could be above the atmosphere, this is generally an expensive endeavor. Instead, astronomers and engineers devised a system where a laser is mounted on a telescope to project an artificial "star" onto the sky. A sensor then observes how the atmosphere deforms this "star" on millisecond timescales and sends commands to the mirror segments of the telescope. These commands drive small motors, which rapidly move each mirror segment, changing the shape of the telescope. This process allows the telescope's geometry to correct for the way the atmosphere distorts a celestial object's image, effectively subtracting out these distortions. This system, called *adaptive optics*, in essence eliminates nearly the entire effect of the atmosphere above the telescope. It's almost as if the telescope is in space!

7.2.5 Space Telescopes

Sometimes, however, it is absolutely necessary to get a telescope above the Earth's atmosphere. There are two reasons for this:

1. Resolution: While adaptive optics can remove much of the effects of the atmosphere, not all can be eliminated—and not every observatory has such a system in place.
2. Absorption: While our atmosphere transmits light in the visible portion of the spectrum, allowing us to see the Sun, Moon, and stars, it is actually quite opaque to other forms of light. **Figure 7.9** shows how the opacity of the Earth's atmosphere changes with wavelength.

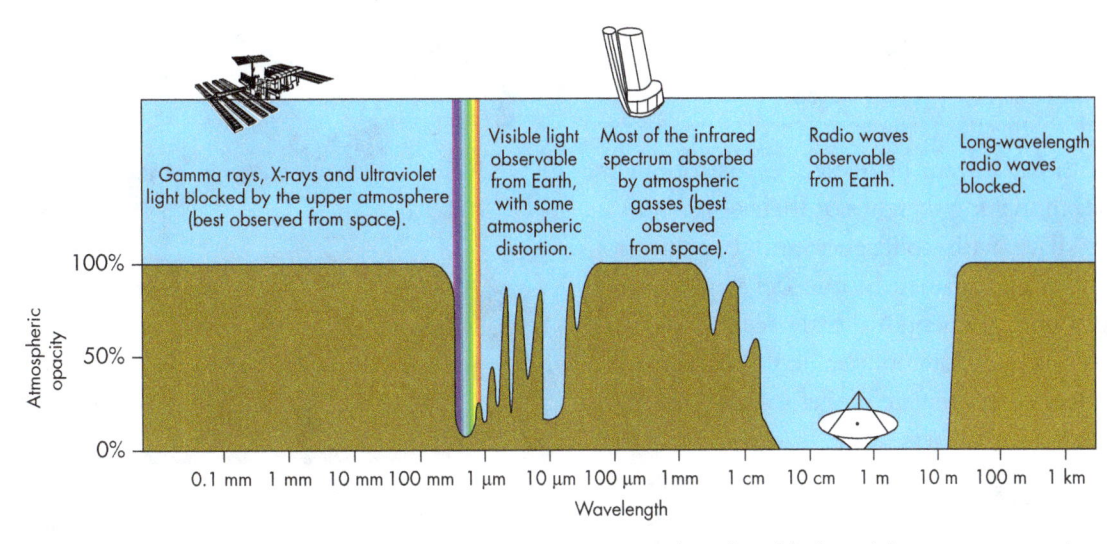

Figure 7.9. The Earth's atmosphere is opaque to some wavelengths of light, while transmitting others. On the surface we can receive visible light, some infrared, and a large portion of radio wave radiation. The atmosphere effectively absorbs gamma rays, X-rays, and much of the UV and infrared. To observe these wavelengths, we must use telescopes in space.

NASA / Public Domain

As you can see, on the short wavelength end it is completely opaque (100% opacity), meaning that gamma rays and X-rays are (thankfully) absorbed and do not reach the ground. Ultraviolet light can partially penetrate with increasing success the closer it is to the violet region (referred to as the near-UV). At longer wavelengths, infrared light is also partially able to penetrate, although water vapor in our atmosphere does an excellent job at absorbing much of it. Finally, while microwaves are mostly absorbed by the atmosphere, many radio waves can easily penetrate. For this reason, astronomers are able to build radio telescopes on the ground. Fortunately for radio astronomers, radio waves can even penetrate clouds, meaning that they can collect data even in bad weather.

On the other hand, if an astronomer is interested in studying objects in wavelengths other than visible light and radio, he or she must utilize telescopes in space. These instruments are above the atmosphere, avoiding both resolution degradation and absorption. It is expensive to put telescopes up in space and a cost-benefit analysis must be done before any space telescope can be approved. In the future, as the private sector becomes more active in space travel and exploration, it may become cheaper and easier to launch and use space-based telescopes.

7.3 Astronomical imaging devices

With the advent of telescopes, astronomers began making sketches of their observations to share with colleagues and benefactors. The insight gained by the development of, and improvements to, early telescopes cannot be understated. Galileo demonstrated that the heliocentric model was correct, and two more planets were discovered in our solar system along with additional planetary moons and a handful of smaller objects called asteroids. Comets were studied and tracked. Gaseous nebulae were discovered, as were great clusters of stars.

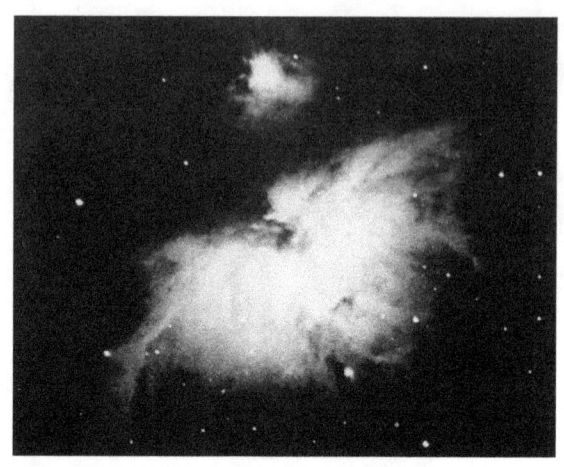

Figure 7.10. An 1883 photograph of the Orion Nebula, which revealed for the first time structure and stars that had previously been invisible to the eye.

Andrew Ainslie Common / Public Domain

By recording the motions of the newly discovered moons, astronomers were able to determine the masses of planets like Jupiter and Saturn. Furthermore, by recording the motions of these planets, astronomers derived estimates for the mass of the Sun.

Sketches of telescopic views were helpful in identifying structures and features, but they had one problem: they were too subjective. Two people might have different abilities to perceive detail, and thus produce slightly different sketches. Furthermore, the human brain is sometimes prone to making a person see what they "expect" to see, leading some astronomers to produce sketches that were more in line with what they *hoped* to see than with what was *actually* there. Finally, since the light coming from most celestial objects is fairly low in brightness, the color receptors in the human eye are not activated. For this reason, nearly every object seen through a telescope (even today) takes on a rather dull gray hue.

Eventually, imaging techniques came into existence. Using glass or metal plates covered with a chemical film that was sensitive to light, astronomers could obtain objective portraits of objects viewed through a telescope. By developing films that were sensitive to different colors of light, astronomers could measure the brightness of objects in blue and red wavelengths, thus obtaining the first quantitative measure of color. Finally, and perhaps most importantly, they quickly discovered that by allowing light to fall onto the photosensitive film for extended periods of time, astronomers could acquire images of details and objects that were too dim to be visible to the human eye (**Figure 7.10**).

The development of long-exposure imaging opened up a huge door in astronomy. Previously, one's ability to see fainter detail depended upon building a large telescope. Once the telescope was built, its ability to see faint detail was set. Each subsequent generation of telescope made the previous generation obsolete. Imaging techniques brought smaller telescopes back into the playing field by allowing users the ability to produce similar images as larger telescopes—at the cost of a longer exposure time due to the reduced light-gathering power. Even today, telescopes with apertures of less than one meter in diameter are still producing data for valuable discoveries, despite the existence of telescopes with apertures ten times that size. Additionally, amateur astronomers using small telescopes are able to produce truly astounding images by combining hours upon hours of exposure time using computers.

In the modern era, glass photographic plates and film have given way to electronic light-sensitive CCDs. These devices create an image that is much more consistent with the photons that strike it, meaning that the image is a very accurate representation of the actual target. Being a digital image, scientists can measure it easily using computer software and process it for calibration. They can also be combined to create color images and scaled or subtracted to reveal hidden details. In most observatories, sets of filters are available to measure brightness in different regions of the visible spectrum.

7.4 Brightness measurement

When astronomers measure an image, they are often measuring the brightness of an object or feature. This brightness can be measured in two different ways:

1. Physical units: Just as we use official units like meters or feet to specify a length, there exists an official unit for how bright something is. An object's intrinsic brightness, that is, how much energy it is emitting, is called its *luminosity*. The unit for luminosity is the Watt (W); you may be familiar with Watts from household light bulbs. If you think of a light bulb (or a star, for that matter) emitting light, it emits it in all directions—its light distribution is spherical. As you get farther from the light bulb or star, you probably already know intuitively that it looks dimmer. But why is this?

    Look at **Figure 7.11**. Suppose we have three shells, like those illustrated, centered on a star. We could draw a square on the surface of the first shell and measure how much light from the star is passing through that square. As the light rays that pass

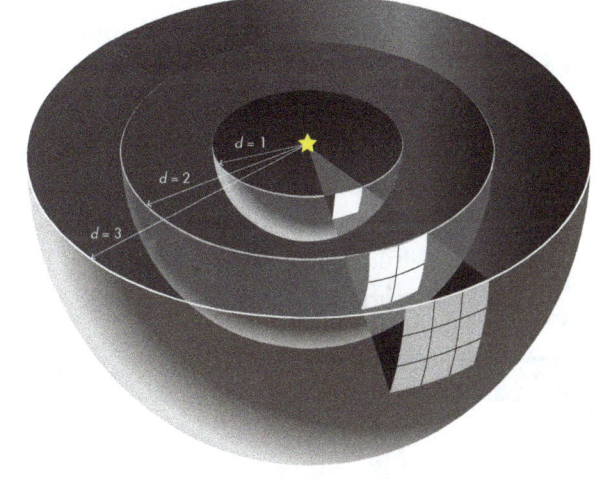

Figure 7.11. As you get farther from a source of light (the star in this figure), the amount of light it emits remains the same but less and less of it enters your eye because the light rays are diverging. Therefore, the star appears dimmer with increasing distance. The apparent brightness drops proportionally to the distance squared—if the distance from the star doubles, its apparent brightness decreases by a factor of 2^2, or 4.

through that square travel onward, they are gradually diverging. Eventually they pass through the second shell, which has a radius equal to twice that of the first shell. In the diagram, you can see that this light now passes through an area on the second shell equal to four times the area of the square it passed through on the first shell. The radius (distance from the star) has doubled, and the light has been smeared out over an area four times as large. Each of the smaller squares shown only contains 1/4 the original light. In other words, the distance doubled, and the *apparent brightness* has dropped by that factor of two squared ($2^2 = 4$).

As that light passes through to the third shell, it now takes up an area equal to nine times that of the first shell's square. The radius has tripled, and the apparent brightness has dropped by a factor of nine ($3^2 = 9$). The apparent brightness of the star at this distance is now 1/9 the apparent brightness from the distance of the first shell.

You can think of your eye as one of those squares. As you back away from a light bulb, the light rays from the bulb are diverging, so fewer of them are ending up in your eye. This makes the light bulb appear dimmer. This principle is called the *inverse square law for light*—the apparent brightness of a star or light bulb drops with the square of the distance. As the distance doubles, the brightness drops by a factor of four. As the distance triples, the brightness drops by a factor of nine. The official units for apparent brightness are Watts per square meter (W/m^2). This reflects the idea that the light from the star (its luminosity, in W) is being smeared out over an area on a spherical shell (in square meters).

2. Conventional units: Before there were any officially defined units or unit systems, the ancient Greeks created a system for specifying how bright stars appeared in the sky. This system was, naturally, based on the human eye's response to the star's light, and they ranked the stars into 6 categories, called apparent magnitudes. The 1st magnitude stars were the brightest, while the 6th magnitude stars were the dimmest.

As time went on, this system was made more precise by ranking stars within any magnitude category using decimals. Hence, a 1.0-magnitude star was brighter than a 1.5-magnitude star. Thus was born the *apparent magnitude system*, shown in **Figure 7.12**.

> **Apparent magnitude**
>
> A ancient system of describing a celestial object's visual brightness. In this system, lower values represent visually brighter objects.

From this perspective, the magnitude system makes sense. However, we typically consider larger values of anything to be "greater" than lower values. This means that the magnitude system is horribly counterintuitive—Pluto, being a 15th-magnitude dwarf planet, is much dimmer than the planet Venus, which has an apparent magnitude of about −4.4 at its brightest. The Sun, by comparison, has an apparent magnitude of −26.7; it is the brightest object in our sky. The naked eye limit is still right around 6th magnitude. Based on this, do you think Pluto is visible to the naked eye?

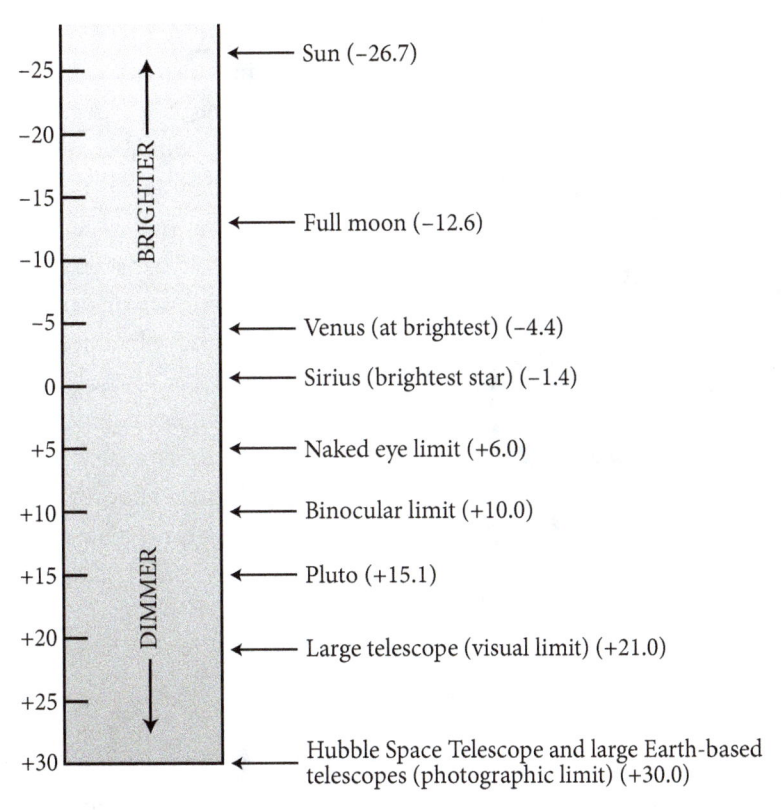

Figure 7.12. In the astronomical magnitude system, small numbers correspond to bright objects. The Sun is the most negative (brightest) on this scale. Large numbers correspond to faint objects.

Just as there exists a physical term for the intrinsic brightness of an object, there is an analog in the magnitude system. Astronomers specify an object's intrinsic brightness as its *absolute magnitude*. However, one key element to this magnitude system is that there exists no special zero point like on a number line—all magnitudes are defined with respect to something else (usually an adopted standard star). To use magnitudes to specify an object's true light output, astronomers have created a special definition based on the object's distance as determined from its parallax. While stellar parallax could not previously be detected by eye, the development of telescopes has made measuring stellar parallax possible in the modern era.

Recall the concept of parallax from Chapter 2. As we orbit the Sun, we observe nearby objects to shift ever so slightly back and forth against the background of more distant stars. This occurs because we are taking different viewing perspectives in January versus July. The angle that the star appears to shift (actually

Parsec

A unit of distance defined such that one parsec is the distance of an object with a parallax angle of one arcsecond. In these units, 1 pc = 3.26 light-years.

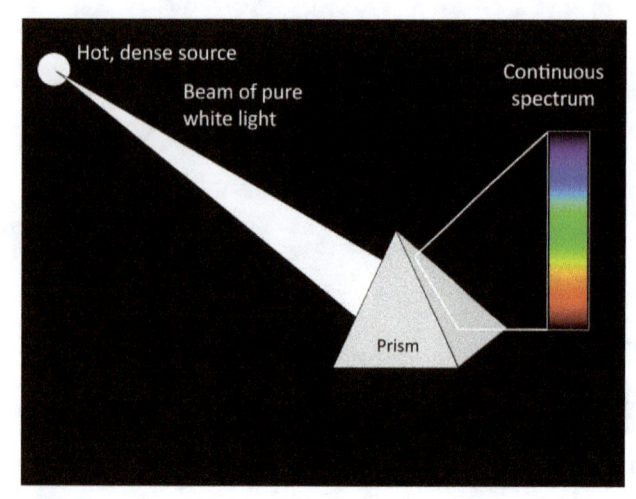

Figure 7.13. Light from a hot, dense source produces a continuous spectrum.

half that angle) is referred to as the parallax angle. Astronomers define a distance unit called the *parsec* *(pc)*, where an object at a distance of one parsec has, by definition, a parallax angle of one arcsecond (1/3600th of a degree). If a star has a parallax angle of two arcseconds, it is at a distance of 1/2 pc. Conversely, if a star has a parallax angle of 1/2 arcsecond, it is at a distance of 2 pc.

To define the absolute magnitude, astronomers say that a star's absolute magnitude *(M)* is equal to the apparent magnitude *(m)* it would have if viewed from a distance of 10 parsecs. With this method, the distance factor is eliminated and all objects are now at an equivalent "viewing distance"; if one star appears brighter than another star from 10 parsecs away, it really is more luminous.

7.5 Spectroscopy

While imaging and brightness measurements in various wavelength regions have provided astronomers with a large amount of information, the study of spectra has opened the floodgates.

You learned in Chapter 6 that atoms absorb and emit light when electrons gain or lose energy. If a photon has an energy equivalent to one of the allowed transitions of an electron within an atom, then the electron can absorb that photon and jump to a higher energy state. The electron will then eventually transition back downward toward the ground state, giving up that gained energy in the process by emitting one or more photons along the way. But in what astrophysical circumstances would we expect to observe these phenomena?

In order for an atom, or a cloud of atoms, to be able to absorb light at all, the atoms must have electrons and those electrons must have relatively low energy—they must be at or near the ground state. If the electrons were all in excited states, there would be fewer opportunities for "upward mobility" to higher energy levels, thus limiting their ability to absorb light. If the atoms have electrons in the ground state, then we would consider this gas cloud to be relatively cool. This cool gas is all set to absorb light … it just needs some light to absorb.

Along comes a beam of pure white light. We will consider it "pure" in that it is a continuous spectrum of light, without any features in the spectrum, as shown in **Figure 7.13**. You learned in Chapter 6 that a continuous spectrum comes from a source that is both hot and dense. In astronomy, this is generally the core of a star. Nuclear reactions in a star's core produce photons of specific energies, but these photon energies quickly become randomized through

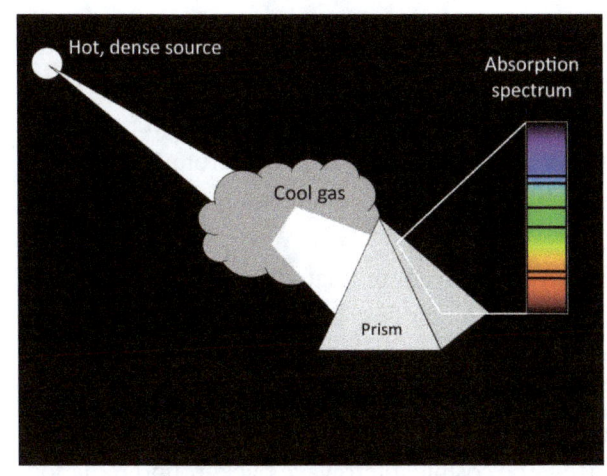

Figure 7.14. Atoms in a cool gas cloud absorb light at specific wavelengths, producing an absorption pattern in the spectrum of the hot, dense source in the background.

collisions with free particles as they work their way outward from the stellar core to the surface. If it were possible to obtain a spectrum of light from within a star's interior, it would appear continuous.

Since this beam is white, it contains radiation at all wavelengths. This means that when it passes through our cool gas cloud, it is highly likely that photons will be present with energies that correspond to allowed energy transitions within the atoms in the gas. As these photons interact with the atoms, they are absorbed. The rest of the light passes through the gas unhindered.

Figure 7.14 shows the resulting spectrum—an absorption spectrum. Remember that this is the spectrum of the hot, dense source when viewed *through* the cool gas cloud. It is not the spectrum of the cloud itself, but rather an indication of the cloud's impact on the spectrum. The light that was absorbed by the atoms has been removed from the beam, while the rest of the light has passed through to be detected. Where light was removed, dark absorption lines appear. The position of these lines correspond to the energy transitions in the atoms within the gas cloud, and thus serve as an identifying "fingerprint" for those atoms.

The atoms in the gas have now been energized. A cloud of energized gas is generally considered to be relatively warm. As you have learned, the electrons that absorbed the light will eventually re-emit it. It is very unlikely that it will be re-emitted in the same direction the light was originally traveling; it is usually re-emitted in a random direction. As the electrons transition down to the ground state, they emit light at the same wavelengths that were originally absorbed. If you could take a spectrum of just the light being emitted by the gas cloud, you would see only light at those specific wavelengths—an emission spectrum (**Figure 7.15**).

In a star, the cooler outer atmosphere (the "surface") acts as the cool gas cloud in this illustration. The hot, dense core produces a continuous spectrum of light that travels outward through the star. As the

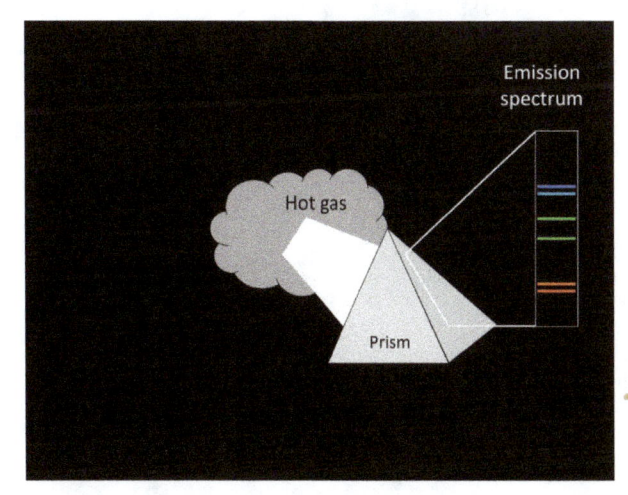

Figure 7.15. A hot gas cloud, or a gas cloud that has been energized by absorbing light, will emit light at specific wavelengths. This produces an emission spectrum.

light passes through the cooler gas at the surface, some light is absorbed while the rest continues onward and outward. Thus, the spectrum of a typical star contains a broad continuous shape with absorption lines transposed on it. Stars generally produce absorption spectra. In some settings, stars are embedded within clouds of gas called nebulae. The light from the stars can energize the gas in the nebula, causing it to emit light. These sorts of emission nebulae produce emission spectra.

The principles and techniques of spectroscopy might seem unrelated to the topic of the big bang, so you may be wondering why we have spent so much time looking at them. As you will learn in the next chapter, while the big bang model was initially a hypothesis proposed to try to explain where the Universe came from and why it is behaving the way it appears, one of the strongest pieces of evidence in support of the model involves spectroscopy. We will refer back to this topic when we look at this piece of evidence—the cosmic microwave background radiation.

CHAPTER EIGHT

Edwin Hubble, arguably the greatest observational astronomer of the 20th century. Hubble's discoveries include proving that the Milky Way is one of many individual galaxies and that the Universe is expanding.

The Expanding Universe

Learning Objectives

In this chapter you will learn:
- » To describe the Doppler effect
- » How Cepheid variable stars can be used to estimate distance
- » How Edwin Hubble resolved the "Great Debate"
- » Why astronomers believe the Universe is expanding

Key Words
- Hydrostatic equilibrium
- Variable star
- Redshift
- Blueshift
- Hubble's Law

8.1 "The Great Debate"

As the 19th century drew to a close, a remarkable number of physical insights had been realized. Scientists were growing extremely confident that much of what could be learned had already been learned. Discussions grew about what else could possibly be yet undiscovered. Science was in an interesting period of certainty and uncertainty. Little did scientists of the time know that some tremendous discoveries were just over the horizon.

The opening of the 20th century brought a flurry of activity: the structure of the atom was being deduced and Albert Einstein was publishing his work on special and general relativity.

While the world of physics was experiencing tremendous productivity, astronomers were entrenched in what would come to be known as "The Great Debate."

Astronomers had been familiar with the Milky Way since ancient times. Galileo had used his telescope to demonstrate that the familiar hazy stripe through the night sky was actually composed of countless stars, too dim and distant to resolve by eye. As telescopes had grown in size, many astronomers used these instruments to attempt to map out the distribution of stars in space as a means of attempting to study the structure of the Milky Way and our location within it. During this process, other objects were uncovered that provoked some interest.

Telescopic observations had revealed a myriad of interesting objects. Structures like gassy nebulae, such as the Crab Nebula shown in **Figure 8.1a**, led astronomers to wonder about the nature of these gas clouds and whether the haziness they saw was truly gas or simply unresolved stars. Star clusters like Omega Centauri, shown in **Figure 8.1b**, motivated astronomers to begin looking for other such massive populations of stars and to map their distributions. However, the most puzzling structures discovered during the 18th century were the oddly shaped "spiral nebulae"—pinwheel-shaped structures that appeared hazy like nebulae but had a significant amount of substructure to them. When considering objects like M101, shown in **Figure 8.1c**, astronomers began to discuss whether or not these objects were actually within the Milky Way.

William Herschel had attempted to measure the structure and size of the Milky Way in 1785. He did so by using his own telescope, which had a focal length of 20 feet and an aperture of about 19 inches, to survey the sky and count up the stars he saw in any given direction. He operated with two fundamental assumptions:

1. Stars are uniformly distributed throughout the Milky Way
2. His telescope was capable of seeing every star out to the very edge of the Milky Way

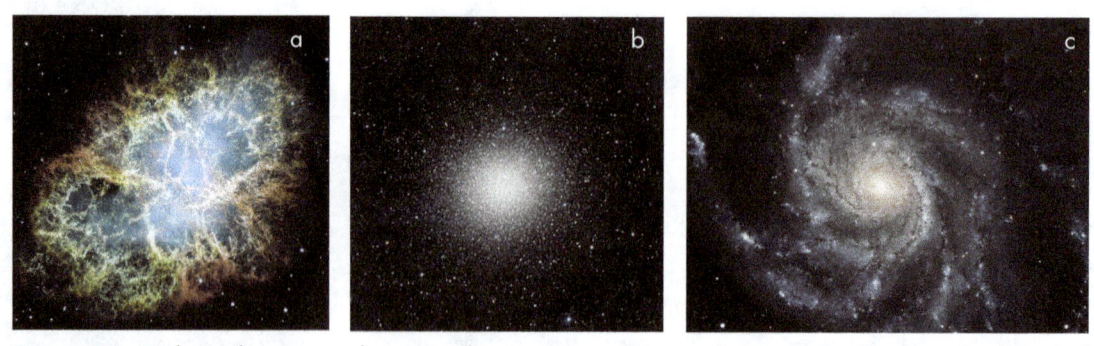

Figure 8.1. Eighteenth-century telescopic observations were revealing a wide diversity of celestial objects, including nebulae (a), star clusters (b), and spiral galaxies (c). Spiral galaxies were known as spiral nebulae at the time.

(a) NASA, ESA, J. Hester, and A. Loll / Public Domain (b) Copyright © 2011 by ESO, INAF-VST, OmegaCAM, A. Grado/INAF-Capodimonte Observatory / CC BY 3.0 (c) NASA, ESA, K. Kuntz (JHU), F. Bresolin (University of Hawaii), J. Trauger (Jet Propulsion Lab), J. Mould (NOAO), Y.-H. Chu (University of Illinois, Urbana), and STScI / Public Domain

Consider each of these assumptions individually. Do you think they were good assumptions or not? Sometimes scientists are faced with questions that are either not yet exactly solvable or are actually impossible to solve exactly. But this fact does not and should not discourage a scientist from pursuing it as much as is possible. However, to do so often requires a set of assumptions or initial conditions to set the stage. These assumptions can then be tweaked later if the results do not match observations.

Herschel knew of the existence of star clusters, so assuming a uniform distribution may not have been warranted. On the other hand, he also knew that many stars were not in clusters so it was simplest to assume that those stars were distributed uniformly. Finally, his assumption that his instrument was capable of seeing all that could be seen was perhaps a bit brazen. He would later recognize that every larger telescope that was built still revealed a whitish haze in the background, indicating the presence of even more distant background stars in the Milky Way. All things considered, and given that such an extensive project had never before been undertaken, he can be forgiven for having made a couple assumptions that made the task more feasible.

Armed with his telescope, and with his sister Caroline as a collaborator, Herschel set about pointing his telescope around the sky and recording counts of all the stars he could see. He then reasoned, based on his second assumption, that the number of stars he could see correlated with the physical extent in that direction. If he saw a lot of stars, then the Milky Way must extend to a great distance. If he saw few stars, then the edge must be fairly nearby. With these numbers in hand, he produced the first map of the Milky Way, shown in **Figure 8.2**. According to his census, the Sun lay somewhere near the center of the Milky Way, which had a shape resembling a disk, and the entire distribution had a physical extent of approximately 10,000 light-years.

What Herschel failed to realize, and what took nearly 150 years for astronomers to fully grasp, was that the Milky Way's disk contains a large amount of dust. This dust obscures our view of distant objects, preventing a complete exploration of the Milky Way in visible light. Observers now turn to infrared light to view beyond this limit, collected by space telescopes or ground-based

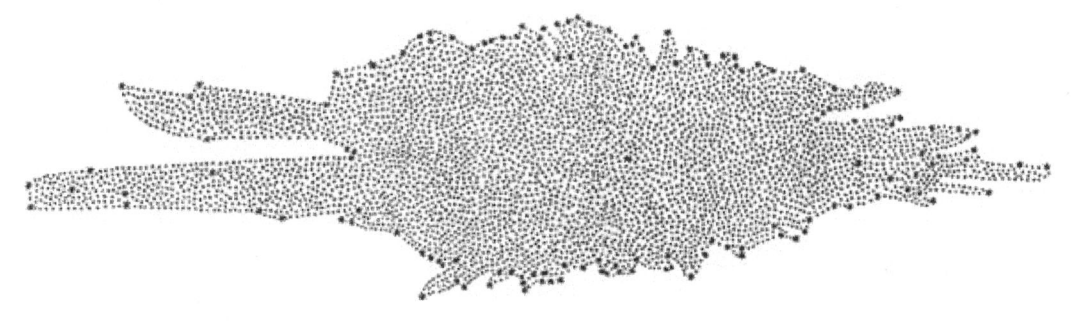

Figure 8.2. The map of the Milky Way created by William Herschel in 1785. Assuming that all stars were distributed uniformly, his map revealed a disk-shaped population of stars with the Sun near the center (dark star near middle). While we now know this to be incorrect, this was the first systematic attempt at mapping a large-scale structure in space.

Caroline Herschel / Public Domain

telescopes on the highest mountaintops above much of the Earth's atmospheric water vapor. The longer wavelength of infrared light allows it to pass through dust with greater ease. To Herschel's eyes, what appeared to be the "edge" of the Milky Way was simply just the limit to which visible light was able to be transmitted through the Galactic dust.

In 1918, Harlow Shapley measured the positions and distribution of globular star clusters (recall **Figure 8.1b**) throughout the Milky Way. Approximately 150 of these clusters exist in our galaxy, and Shapley started with the assumption that their spatial distribution was symmetric about the Galactic center. In order to determine their absolute positions within the Galaxy, Shapley needed to know accurate distances. To measure this, he used variable stars.

Figure 8.3. A variable star has a rhythmic variation to its luminosity, and therefore its apparent brightness. Such a variation, when plotted over time, is called a *light curve*. The pulsation period of a variable star can be determined from light curves like these.

> **Hydrostatic equilibrium**
>
> The state of balance established in most stars where the inward gravitational force is balanced by an outward force produced by energy generated through nuclear fusion inside the star.

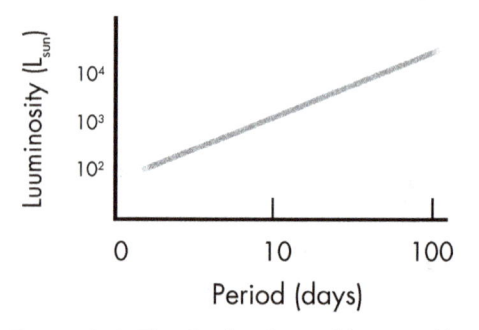

Figure 8.4. The Cepheid variable period-luminosity relationship, as discovered by Henrietta Leavitt in 1912. This relationship between a Cepheid's luminosity and its pulsation period allows astronomers to easily determine the intrinsic brightness for these stars, and then use the inverse square law to find their distance.

Within a star, there are two dominant forces at work. The energy generated by nuclear fusion reactions in the core attempts to push outward, while the gravitational force due to the star's mass is constantly pulling inward. Generally, these two forces balance each other out and the star exists in a state of *hydrostatic equilibrium*. However, as a star ages and evolves, its interior goes through phases where this equilibrium is not in place. At certain points late in a star's evolution, the interaction between thermal pressure outward and gravitational pressure inward gets out of sync and the star begins to pulsate, rhythmically growing and shrinking in size. This change in size also corresponds to a change in brightness, shown in **Figure 8.3**. The object is then known as a *variable star*.

Earlier, in 1912, Henrietta Leavitt published her results regarding a specific type of variable star called a Cepheid variable. She discovered that there was a distinct and uniform relationship between the pulsation period of these stars—the time it takes for the star to go from bright to dim to bright—and their intrinsic luminosity. **Figure 8.4** illustrates this relationship, which makes determining the luminosity of a Cepheid variable quite easy. One simply produces a light curve similar to that shown in **Figure 8.3**, reads off the pulsation period, then uses **Figure 8.4** to estimate the luminosity that corresponds to that period.

The real value in this relationship is what comes next. Once the luminosity is known, and the average apparent brightness has been measured from the light curve, the distance can be easily calculated using the inverse square law for light in mathematical form:

$$b = \frac{L}{4\pi d^2}$$

In this equation, b is the apparent brightness and L is the luminosity. The distance, d, can be determined algebraically using these two quantities. Your brain actually gauges distance using this relationship all the time—it is built into your intuition. The equation above is simply useful from a mathematical perspective.

Harlow Shapley identified variable stars in the globular clusters he was observing and assumed they were Cepheid variables*. Using the relationships described above, he obtained distance estimates for the clusters and produced a diagram illustrating their distribution throughout the Milky Way, shown in **Figure 8.5**. Using his assumption that their distribution was symmetric about the Galactic center, he noticed that this point at the center of the star cluster distribution was located a great distance away from the Sun. This meant that the Sun was *not* located at the center of the Milky Way, as Herschel had claimed.

This was a pivotal result that actually built upon a major change initiated by Copernicus 400 years earlier. Up until the Renaissance era, it had been believed that Earth was at the center of the Universe. The Copernican heliocentric model shifted that center from Earth to the Sun. It was then believed that the Sun was at the center of the Universe. Shapley had just shown that this was not true either, moving the Sun to a seemingly arbitrary location within the Milky Way.

The other result that came out of Shapley's work was that he calculated the Milky Way's size to be about ten times the size Herschel had originally reported. Suddenly, the Universe was a very big place!

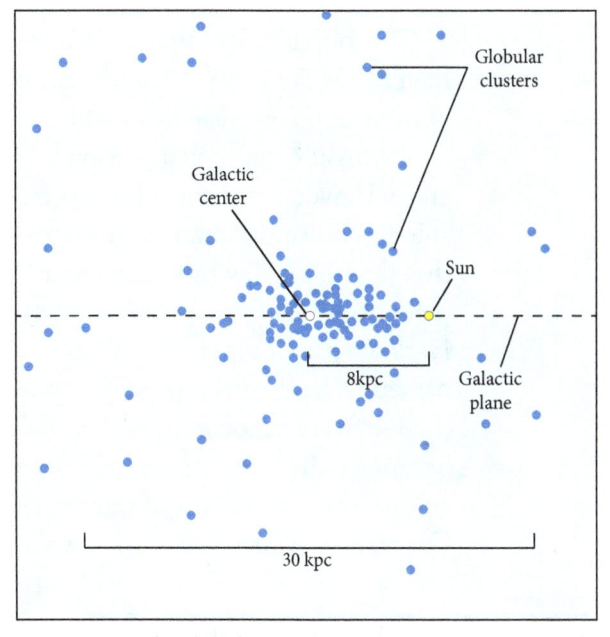

Figure 8.5. The distribution of globular clusters throughout the Milky Way, as determined by Harlow Shapley in 1918. This showed that the Sun was not located at the center of the Milky Way, and that the Milky Way was a much larger place than had been originally believed.

> **Variable star**
>
> A star that changes its brightness due to changes in its size.

---

* While Shapley was incorrect in his assumption—the stars were actually a different type of variable star called *RR Lyrae* stars—his main results remain true even if the distances he calculated were a bit high.

This development brought up some very big questions—was there more to the Universe than the Milky Way? What was the overall structure of the Milky Way? What was the nature of some of the nebulae that had been discovered?

Discussion and debate ensued. One side maintained that the Milky Way comprised the entire Universe and contained everything that could be seen. The other side claimed that the spiral nebulae that had been discovered were actually "island universes" of their own and that the Milky Way probably looked very similar. No evidence existed to support either side definitively, but this did not stop astronomers from holding an "official" debate in 1920, with Harlow Shapley representing those who believed in the former and another noted astronomer named Heber Curtis representing the latter.

Both were famous within their field, and both were adept public speakers. Shapley penned his approach in a more general way, easily accessible to even the non-astronomers in the audience. However, his lack of depth meant he made few successful points to support his thesis. Curtis, on the other hand, took a professional approach, creating typed slides and presenting in a technical way that was rich in detail. Despite this, he was still unable to successfully prove his side. While most generally agreed that Curtis won "The Great Debate," the question was still largely unresolved. Vital observational evidence supporting Curtis' case was still missing.

But not for long.

8.2 Edwin Hubble

Edwin Hubble (shown in the chapter opening) was born in Missouri in 1889 and was exposed to astronomy at the age of eight when his grandfather bought him a telescope for his birthday. Fascinated by the subject, Hubble earned a scholarship to the University of Chicago where he studied law and physics. After continuing his studies in England, Hubble returned to the United States and eventually obtained a research position at the Mt. Wilson Observatory in California. This observatory was home to the mammoth 100-inch Hooker Telescope, the world's largest telescope at the time (**Figure 8.6**).

He was able to get time using the instrument and in 1923 made 40-minute exposures of M31, the Andromeda Nebula (as it was then called) on consecutive nights. There were several objects that struck his interest, and one of these

Figure 8.6. Hooker Telescope at the Mount Wilson Observatory in Los Angeles County, California that was used by Edwin Hubble.

Copyright © 2005 by Ken Spencer / CC BY-SA 3.0.

objects was a Cepheid variable. It appeared on some of his photographic plates, but not on others, revealing the fact that it was pulsating. He excitedly marked it on the plate, knowing in his mind that such a variable star had never before been discovered in a spiral nebula.

Measuring the Cepheid's brightness and period, Hubble crunched the numbers to arrive at the distance to the Andromeda Nebula. His result was astonishing: 900,000 light-years. This distance was roughly nine times the known size of the Milky Way! Certainly this meant that the Andromeda "Nebula" was no nebula, but was in fact an island universe of its own and not located within the Milky Way. Knowing the impact that this result would have on the scientific community, Hubble carefully made more observations to verify his conclusion. He formally announced his results in 1924, to great applause. He had resolved the Great Debate. Hubble's name and accomplishment quickly spread throughout the country, and beyond.

His discovery opened up the entire Universe as a collection of "island universes"—relabeled *galaxies*. Galaxies were found to come in two different shapes: spiral and elliptical. While embracing his newfound celebrity, Hubble spent the next couple years investigating the nature of galaxy morphology, hoping to identify an evolutionary sequence that would explain their shapes. In the meantime, he acquired more data on distances to these galaxies, slowly building up a catalog.

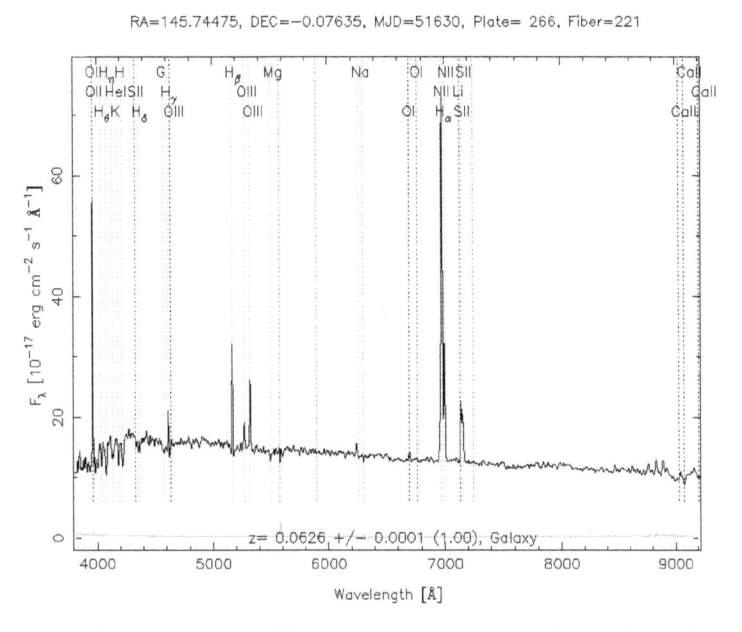

Figure 8.7. A spectrum of a spiral galaxy. The continuum represents the total light from the existing stellar populations in the galaxy, while the tall emission lines are the result of hot gas heated up by star formation.

Source: Sloan Digital Sky Survey

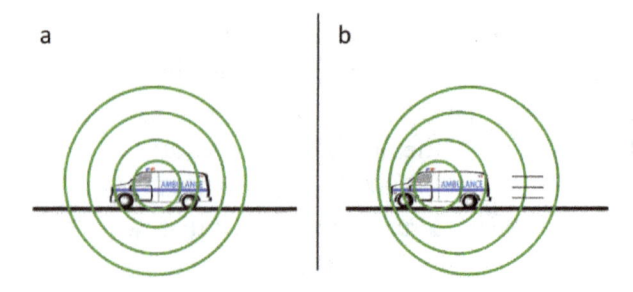

Figure 8.8. The Doppler effect is the shift in the *detected* wavelength of a sound or light wave due to the relative motion of an object with respect to the observer along the line of sight. The ambulance driver hears the siren at the same pitch as always, since he is at rest with respect to the ambulance. Someone in front of the moving ambulance hears the siren at a higher pitch, while someone behind the ambulance hears the siren at a lower pitch.

8.3 The "Stretching" Universe

Roughly a decade earlier, an astronomer named Vesto Slipher had made a different, yet still remarkable, observation. Galaxy spectra look similar to stellar spectra in that they have a continuum (since their light is the sum of all the light from all the stars they contain) and absorption lines. Some galaxy spectra have emission lines as well, as you can see in **Figure 8.7**, signifying the presence of star formation.

Slipher had been measuring the observed wavelengths of the spectral features from several galaxies, though they were still called "nebulae" at the time. He noticed that the spectral features were shifted in wavelength from where they would normally be if the galaxy was at rest with respect to the observer. This wavelength shifting is called the Doppler effect.

The Doppler effect was first predicted by Christian Doppler in 1842. His claim was that the movement of an object would affect the sound waves it emitted, according to the perspective of the listener. **Figure 8.8** helps to illustrate this concept. If an ambulance is stationary as it blares its sirens, then the sound waves will "sound" the same to all observers—the detected wavelengths are the same regardless of where you are located with respect to the ambulance. However, if the ambulance is moving then the sound waves will seem to be compressed in the direction of its motion. What this means is that an observer standing in front of the ambulance will hear the siren at a higher pitch (a higher frequency) than they would if the ambulance was stationary. On the other hand, an observer standing behind the moving ambulance, watching it drive away, will hear the siren at a lower pitch because the wavelengths are being stretched out longer. It is worth mentioning that the driver notices nothing strange, since he is at rest with respect to the siren attached to his ambulance.

Light emitted by a moving object behaves the same way. If an absorption line in the spectrum of an object at rest has a wavelength of 656 nm, then as that object begins moving toward an observer, he or she will observe that line's wavelength to shift to a shorter wavelength—a *blueshift*. The light waves are being compressed in the direction of the object's motion, from the observer's perspective, so every feature in the spectrum appears to shift toward the blue end of the spectrum. On the other hand, if the object is moving away from the observer, he or she will observe the feature shift to a longer wavelength—a *redshift*, shown in **Figure 8.9**. The principle is the same as in **Figure 8.8**.

> **Blueshift**
>
> An observed shifting of an object's spectrum to shorter wavelengths (higher energy).

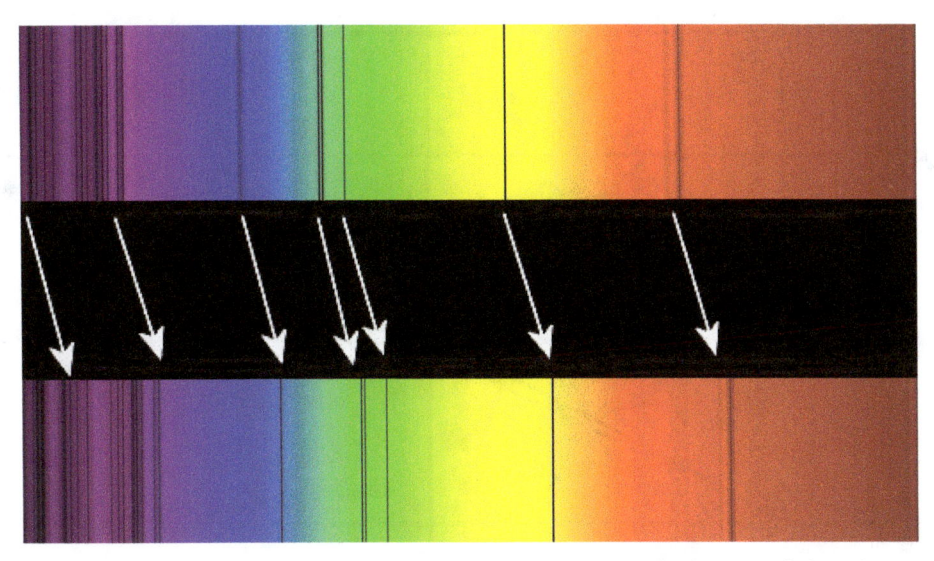

Figure 8.9. The Doppler effect for light. A moving object's spectrum appears to shift to shorter or longer wavelengths depending on whether the object is moving toward or away from the observer, respectively. The shift above shows a spectrum shifted toward the red end of the spectrum: a redshift. Is this object moving toward or away from you?

Copyright © 2009 by Georg Wiora / Wikimedia Commons / CC BY-SA 2.5

Christian Doppler showed that the degree to which this apparent shift occurs is directly related to the velocity of the object. This means that an object moving slowly away from the observer would have a small redshift, whereas an object moving quickly away from the observer would have a large redshift. Based on the amount of shift the spectrum reveals, the observer can determine how fast the object is moving.

> **Redshift**
> An observed shifting of an object's spectrum to longer wavelengths (lower energy).

Vesto Slipher observed that the spectra of galaxies were also shifted, and by quite a bit. Slipher measured a blueshift in the spectrum of the Andromeda galaxy that corresponded to a velocity of about 300 km/s. Other galaxies (or nebulae, as they were still believed to be at that time) showed redshifted spectra, with velocities up to 1,000 km/s. Since these velocities were measured during the time when many astronomers believed that these objects were within the confines of the Milky Way, all were astounded that these objects could be moving at such high speeds—much faster than had ever been seen before. It was even odder that most of the spectra were redshifted. These strange nebulae seemed to preferentially be moving away from the Earth.

When Edwin Hubble heard about Slipher's collection of galaxy spectra, and learned that the vast majority of these spectra were highly redshifted, he was intrigued. Feeling that it was up to him to solve the mystery, Hubble and his observing partner, Milton Humason, began collecting images and spectra for nearly fifty galaxies. After having determined both velocities

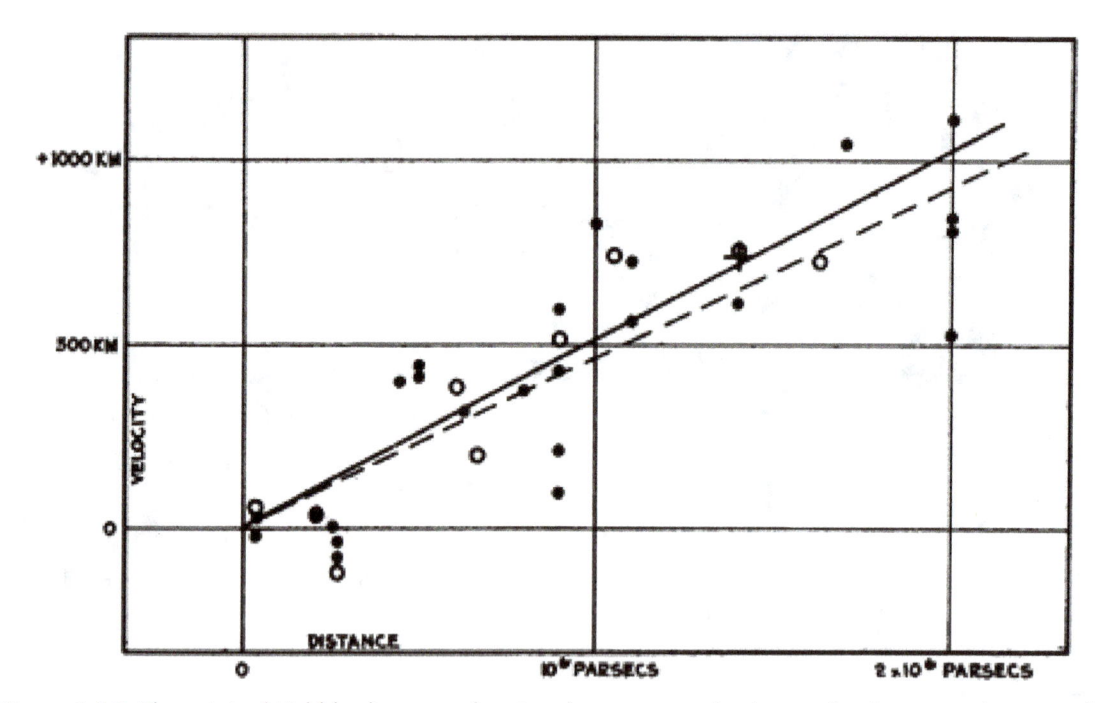

Figure 8.10. The original Hubble diagram, showing the recessional velocity of galaxies on the vertical axis and distance on the horizontal axis. The line drawn indicates a direct relationship between the two quantities—the observed recessional speed of a galaxy is proportional to its distance.

Source: Edwin Hubble, Proceedings of the National Academy of Science, vol. 15, no. 3.

and distances for their sample, Hubble produced a graph plotting these velocities versus the distance to each galaxy. His result is shown in **Figure 8.10**.

The data points indicated that more-distant galaxies were receding from Earth faster than nearby galaxies. Hubble even drew in a line among the data points to illustrate and quantitatively describe this direct relationship. If Galaxy A is twice as far away as Galaxy B, then Galaxy A is moving twice as fast as Galaxy B. The equation for this line became known as Hubble's Law:

$$V = H_0 d$$

The parameter H_0 (pronounced "H-naught"), corresponding to the slope of the line, came to be known as the Hubble Constant. To his credit, Hubble originally labeled this slope with a simple letter K.

At first blush, this might suggest that the Milky Way, Earth's home galaxy, is at the center of a Universe where everything is moving away from it. However, it was quickly realized that what this in fact illustrated was an

> **Hubble's Law**
>
> The observation that the more distant a galaxy appears, the faster it appears to be receding from us.

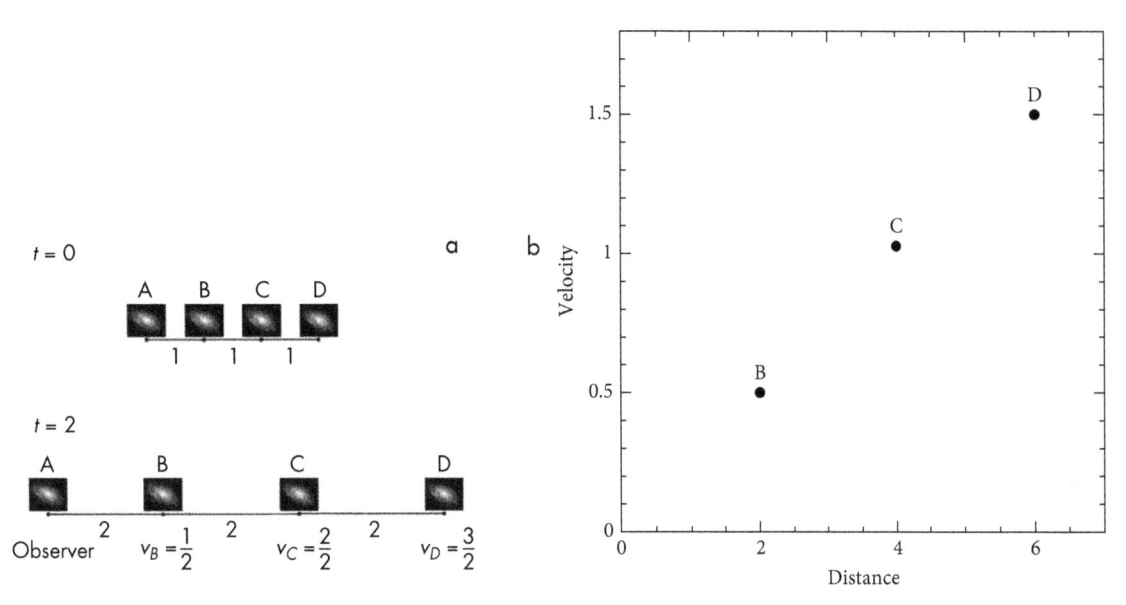

Figure 8.11. (a) Four galaxies initially separated by a distance of 1 move apart until their distance has doubled over a time period of 2. The apparent recessional velocities, according to Galaxy A, are calculated as the change in distance divided by the time elapsed. (b) According to the observer in Galaxy A, the apparent recessional velocities of each galaxy increase with distance.

inherent motion of nearly every galaxy away from nearly every other galaxy. To understand this, let's look at an illustration.

Look at the situation shown in **Figure 8.11a**, where four galaxies are separated by a distance of 1 (in some arbitrary unit) at the beginning of our simulation. You, the observer, are located in Galaxy A. We let two "time steps" elapse as the distance between each galaxy grows from 1 to 2. Let's consider how the Hubble diagram would look from Galaxy A.

To calculate the velocity of each galaxy, divide the change in distance by the time elapsed. Galaxy B has moved a distance of 1 unit; it went from a distance of 1 to a distance of 2. It did so over a time of 2. This means that you, located in Galaxy A, would measure its velocity to be 1/2. Galaxy C has moved a distance of 2 units over the same time interval, so its velocity is 2/2 = 1. Galaxy D has moved a distance of 3 units over the time interval, so its velocity is 3/2 = 1½. **Figure 8.11b** shows that these data points, plotted on a Hubble diagram, reproduce what Edwin Hubble saw in his sample of galaxies. Now let's look at the situation from the perspective of Galaxy D, shown in **Figure 8.12a**.

From Galaxy D's perspective, Galaxy C travels a distance of 1 over the time interval of 2 "time steps." Therefore, its velocity is 1/2. Galaxy B travels a distance of 2 over the same time interval, so its velocity is 2/2 = 1. Likewise, Galaxy A travels a distance of 3 over the same time interval, so its velocity is 3/2 = 1½. The Hubble diagram from Galaxy D's perspective, shown in **Figure 8.12b**, is identical to that seen by Galaxy A. Both perspectives report that all galaxies

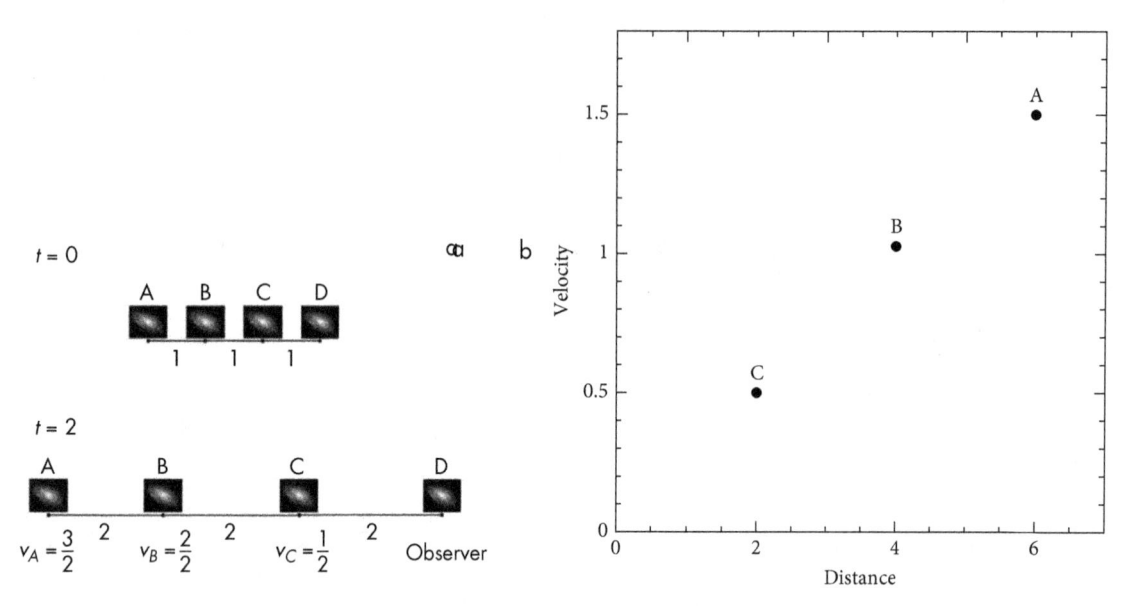

Figure 8.12. (a) Four galaxies initially separated by a distance of 1 move apart until their distance has doubled over a time period of 2. The apparent recessional velocities, according to Galaxy D, are calculated as the change in distance divided by the time elapsed. (b) According to the observer in Galaxy D, the apparent recessional velocities of each galaxy increase with distance, just as is seen from Galaxy A.

appear to be receding away from them, with the recessional velocity increasing with distance. Neither galaxy is in a particularly special location within the Universe—all four galaxies are moving away from one another. There is no special "center".

If this interpretation is true, then this means that in the past all galaxies were originally much closer to one another than they are now. But what could cause this? And are the galaxies all flying away from one another through space, or is something altogether different happening?

To answer this question, we first need to define the redshift parameter, z. The parameter z indicates the degree to which an object's spectrum has been shifted from its rest position due to its relative motion. This parameter is related to the object's velocity by the equation

$$z = \frac{v}{c}$$

Looking at this equation, you can see that the velocities will be given as a fraction of the speed of light. In other words, a redshift of $z = 0.1$ means that the object is receding from Earth at a velocity of $0.1c$; one-tenth the speed of light. A redshift of $z = 0.5$ corresponds to a recessional velocity of half the speed of light.

As it turns out, objects have been observed with redshifts equal to, and greater than, $z = 1.0$. But this means that their apparent recessional velocity is equal to or greater than the speed of light. This might seem confusing, since you may recall from Chapter 5 that the speed of light

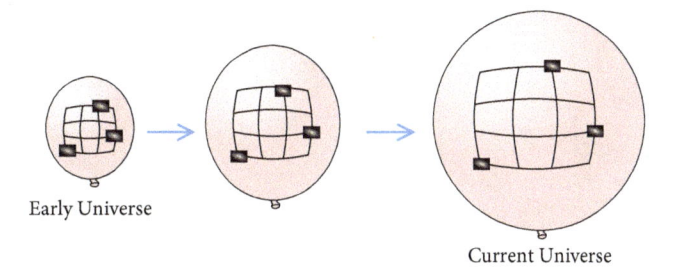

Figure 8.13. As the Universe stretches and expands, it drags galaxies and galaxy clusters along with it. Each galaxy sees every other galaxy moving away from it. No galaxy lies at the "center", just as there exists no "center" to the balloon's surface.

was the "cosmic speed limit"—no material object can travel at or faster than this speed. So what gives?

This speed limit, as it is, applies only to objects moving *through* space. It does not, however, apply to space *itself*. As far as anyone can tell, space is free to move as quickly as it likes. This means that the only way we can explain galaxies that appear to be receding from us faster than the speed of light is if space itself is actually stretching at all points and is pulling each galaxy along with it. **Figure 8.13** attempts to illustrate this by comparing the Universe to the surface of a balloon. As the balloon expands, the surface stretches and the galaxies that are pinned to its surface move along with it.

While this chapter title uses the word "expanding," this is an unfortunate choice of wording that has persisted over the last century. The idea of expansion connotes the image of something growing in size within a larger, pre-existing space—much like the balloon in **Figure 8.13**. However, the Universe is really much more like the fabric of the balloon than the balloon itself. There is no larger, pre-existing space beyond the Universe into which the Universe is expanding. All that exists *is* the Universe. The effect observed by Edwin Hubble is that the fabric of the Universe is stretching out, causing space between galaxy clusters and isolated galaxies to grow, hence the use of the word "stretching" for this section title. The Universe is not expanding out into anything; it is simply growing in size.

Notice that the galaxies themselves do not grow larger. Although the Universe is stretching and growing larger, galaxies themselves are bound together gravitationally. This means that at the current rate of expansion, the force of gravity is still strong enough to keep a galaxy bound up tight. The same goes for stars, planets, and you. The molecular forces holding you together are stronger than the expansion of the Universe. Even galaxies within galaxy clusters are gravitationally bound together within the cluster, so the cluster itself does not grow. Two widely separated galaxy clusters, on the other hand, will be pulled away from one another as the Universe expands.

Let's go back to **Figure 8.13** for a moment. You might have realized that since the galaxy is "pinned" into place, there isn't exactly much motion *through* space taking place. So what, then, is the cause of the Doppler-shifted spectra that Slipher and Hubble observed?

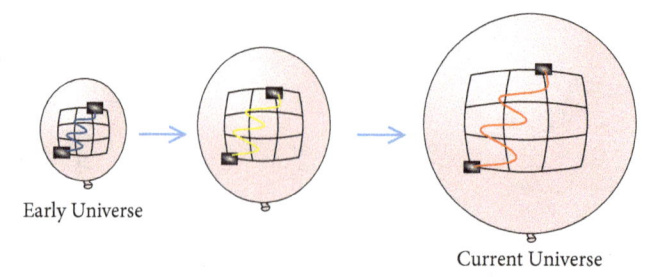

Figure 8.14. As the Universe stretches out the distance between an emitting galaxy and an observing galaxy, the light ray that travels from one to the other is seen to be stretched out by the expansion of space. The observer sees it having been redshifted.

We can once again envision the stretching Universe as the fabric of a balloon, shown in **Figure 8.14**. Light rays that are emitted by stars within a galaxy travel through space toward another galaxy, but as they are traveling, the space between the two galaxies is stretching. Since Galaxy A is being pulled away from Galaxy B, an observer in B will see the light rays from A being stretched out as they travel through space, redshifted to longer wavelengths. The light rays aren't slowing down—they still travel at the speed of light—but from the observer's perspective they are losing energy as space stretches out. This is called the *cosmological redshift*. The cause of the spectral shift is different from the Doppler effect, but the end result is the same.**

Hubble's announcement of an expanding Universe in 1929 was received with a fair amount of skepticism. This occurred for two reasons. The first reason was that there was quite a bit of scatter in his data. Looking at **Figure 8.10**, one can understand why many scientists were not convinced by the line Hubble drew. The data points were widely scattered and their corresponding levels of uncertainty were fairly high. Hubble would need more data, or better data, before his claims could be more substantial.

The second reason his results were questioned had to do with the greatest scientist of the day—Albert Einstein. When Albert Einstein had been developing his General Theory of Relativity, he had devised a rather elegant equation that succinctly described how the Universe was shaped and how it could behave. His mathematical description of spacetime was beautiful, and many physicists rather quickly jumped at the opportunity to see what kinds of universes were allowed by his equation.

The first physicist to propose a solution to Einstein's equations (where a "solution" here means a mathematical representation of a universe that was described by Einstein's relativity equation) was Willem de Sitter. The "de Sitter universe," as it's called, was devoid of matter.

** The Doppler effect, and the cosmological redshift, are both the result of perspective—the observer and the emitter are in different reference frames. How do you think the photon views the Universe as it travels? Do you think it sees its own energy changing?

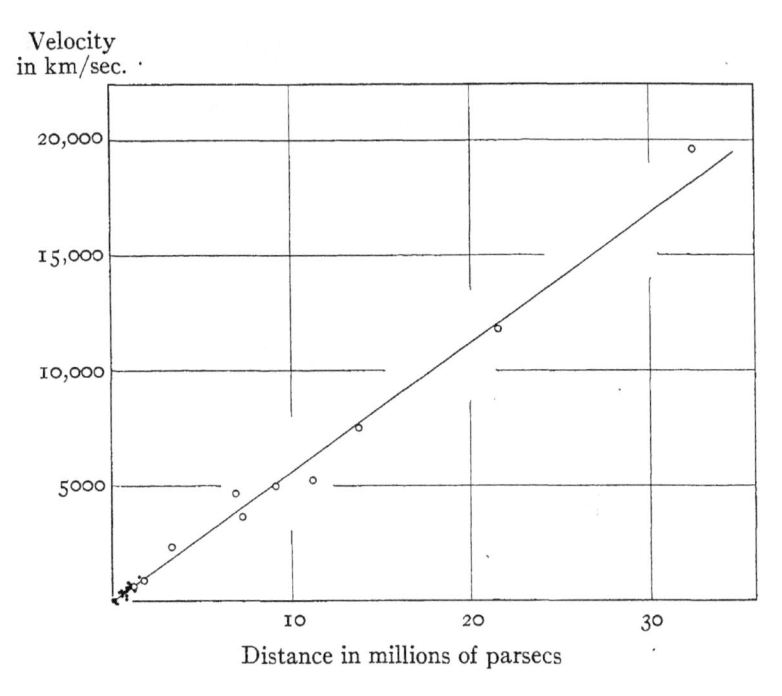

Figure 8.15. Hubble's follow-up publication in 1931 extended his sample to much greater distances. The original data can be seen as the small clump at the lower-left corner.

Source: Edwin Hubble, Astrophysical Journal, vol. 74, p. 43.

However, one interesting element that showed up in the solution was that it was expanding. Additional solutions by physicists Alexander Friedmann and Georges Lemaître also revealed a dynamic, changing universe that could be either expanding or contracting, depending on the initial conditions. An unchanging universe seemed to be unstable.

Albert Einstein was strongly opposed to the notion of an evolving Universe, principally because it reeked of the idea of a "moment of creation." The theological undertones were something he felt ought to be avoided by scientists. He instead favored a Universe that had always existed in a static state. However, he was faced with the fact that, mathematically, his contemporaries were not incorrect. The equations of general relativity allowed a static, eternally unchanging Universe to exist only if it was tuned just right, and even then it was extremely unstable. In order to address what he felt was a physically unrealistic solution to his theory, Einstein inserted a factor called the *cosmological constant* into his general relativity equations. This factor's value could be tuned to produce a static, eternally unchanging Universe. This satisfied both Einstein and many of his colleagues, although both Friedmann and Lemaître protested this modification vigorously.

Hubble, having been largely dismissed by his colleagues, plunged back into his work and steadily accumulated more data. In 1931, he and Humason published a much larger sample of galaxies that extended out to much greater distances (**Figure 8.15**). His original result held up—the recessional velocity was directly proportional to the distance. There seemed to be no denying the observation—the Universe was expanding.

Unfortunately, any good scientific theory must be supported by observational evidence, and Hubble's observations did not support Einstein's cosmological constant. Faced with the reality of an evolving Universe, Einstein was forced to drop his cosmological constant, although he was somewhat comforted by the fact that science had risen to the opportunity to correct him.

8.4 A new paradigm

Hubble's results had produced a monumental shift in how the Universe was to be seen. If galaxies were receding from one another, then at one point they were much closer to one another. In fact, the Universe itself must have been much more compact. Packing all the matter in the Universe into a small region of space would have produced extremely high temperatures. The Universe had had a hot childhood.

While much of the scientific world was surprised by Hubble's results, there was one man who was not. Georges Lemaître, a Belgian astronomer and Catholic priest, had voraciously been working through his solution to Einstein's general relativity equations and had published his hypothesis of an expanding Universe in 1927. He had even derived what came to be known as Hubble's Law and estimated the value of the Hubble constant. However, the fact that it was published in a small Belgian journal, which was not widely read outside of his country, meant that it made no immediate impact on the scientific community. He even discussed his idea with Einstein at a conference in Brussels in 1927, but to no avail. However, once his paper was translated into English following Hubble's discovery, Lemaître's reputation grew. Inspired by an interest in radioactive decay, where an atom can emit a particle and change into a different element, he ultimately proposed that the Universe had initially existed in an unstable state he called the "primeval atom," from which all the matter in the Universe had originated. While the primeval atom idea was eventually modified, Lemaître and Hubble successfully introduced the world to the idea of an evolving Universe that had originated from a hot, dense state.

But what could the Universe have been like in that early time? And why did it suddenly begin expanding? And where did it even come from to begin with? These fundamental questions arose out of the fanfare of Hubble's observations and provided physicists and astronomers with something new to chew on for the next several decades.

CHAPTER NINE

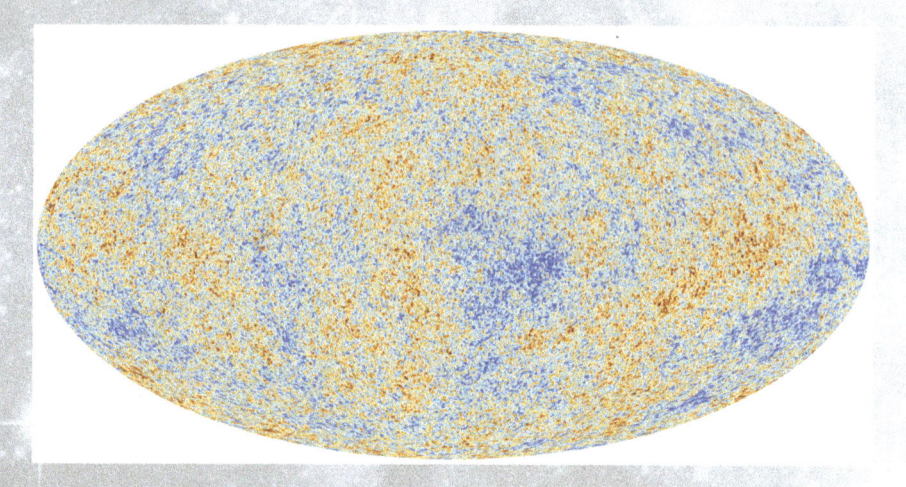

The cosmic microwave background, relic radiation believed to have originated in the distant past some 300,000 years after the Big Bang. Its existence is held to be one of the key pieces of evidence supporting the Big Bang model. The data above come from the Planck mission, a European effort that provided this map of cosmic temperature fluctuations with the highest resolution to date.

ESA and Planck / Public Domain

The Big Bang Model

Learning Objectives

In this chapter you will learn:
- » How opponents of the Big Bang model produced an alternative model
- » Predictions made by the Big Bang model
- » The observational evidence that supports the Big Bang model
- » Key details about each epoch of the Universe's evolution as described by the Big Bang model

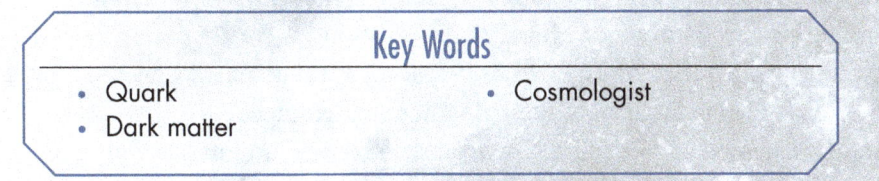

Key Words
- Quark
- Dark matter
- Cosmologist

9.1 A rival model arises

By the early 1930s, astronomy was taking off once again—perhaps as it never had before. Based on the equations of Albert Einstein's theory of general relativity, Georges Lemaître and Alexander Friedmann had proposed that the Universe could not be static but must be evolving. Lemaître had gone so far as to predict the nature of the expanding motion and even estimated the rate of expansion.

Edwin Hubble had resolved the Great Debate, demonstrating that the Milky Way is just one of countless individual galaxies in the Universe, and had shown that the Universe was indeed expanding. Einstein had recanted his static Universe model and publicly endorsed the expanding Universe. Based on Hubble's measurements of the rate of expansion, the Universe appeared to be approximately 2 billion years old, meaning that at some point in the past the Universe as we know it began expanding from a hot, dense state. By the 1940s, the model that explained this had come to be known as the "Big Bang" model, a moniker that served to illustrate how the model explained the creation of the Universe.

There were, however, two immediate problems. The first problem was that rocks on the Earth had already been dated to be older than 2 billion years. How could Earth rocks be older than the Universe? Astronomers dismissed this inconsistency as merely something to be resolved with better observations.

The second problem, though less scientific in nature, became a larger impediment. Despite Hubble's data, it was still the "mainstream" view that the Universe was eternal. Astronomers backing this view did not want to give up this view, especially when the idea of creation was being espoused by Lemaître, an ordained priest. Those who adhered to the static Universe view would not give up without a fight.

While several initial attempts at explaining away the notion of an expanding Universe were offered, they were of little substance. The major contestant model that rose up against the Big Bang model was the brainchild of the very man who coined the phrase "Big Bang"—Fred Hoyle.

Fred Hoyle was born in England and earned his Ph.D. at Cambridge. He was by all accounts a brilliant student. While serving as a radar researcher during World War II, Hoyle met Thomas Gold and Hermann Bondi, and following their service, their friendship and collaboration continued. In 1946, the three men proposed a novel idea that retained the eternal Universe concept while also accounting for Hubble's observations.

Their model, called the *Steady State model*, accepted that galaxies were receding from one another. In order to maintain this eternally, new matter needed to be created in the space left behind by the receding galaxies. Their model assumed an infinite Universe, since only an infinite Universe could expand forever and keep on expanding, and calculated that the amount of matter that needed to be created to maintain this evolving yet unchanging Universe was actually quite small—about one atom per million cubic meters every century.

At the mention of spontaneous creation of matter out of nothing, many physicists cried foul. Matter cannot simply be created out of nothing, they pointed out. Rather than point out that some versions of the Big Bang model also invoked a spontaneous moment of creation, Hoyle proposed the existence of a creation field, referred to as the *C-field* (**Figure 9.1**), which permeated the entire Universe and contributed the energy needed to create particles of matter while also driving the expansion.

Hoyle, Gold, and Bondi had developed what they believed to be a self-consistent model of the Universe that satisfactorily accommodated expansion while also retaining its eternal nature. Now it was simply a matter of making some testable predictions.

One prediction seemed fairly clear: if new matter was being formed in between existing galaxies, then eventually that new matter would form new galaxies in between older galaxies. This means that one might expect the distribution of old galaxies and new galaxies to be fairly well mixed. This should be true nearby within the local Universe as well as at great distances, where we see the Universe as it appeared in the past.

Recall the distance scales of the Universe from Chapter 1. Galaxies are really far away—even the close ones! The Andromeda Galaxy, nearby by astronomical standards, is still 2.5 million light-years away. This means that light emitted by stars in the Andromeda Galaxy takes 2.5 million years to reach us. Therefore, the image of the Andromeda Galaxy we see now represents how it looked 2.5 million years ago. To see what it looks like now, at this very moment, you'll need to wait 2.5 million years. Try to be patient.

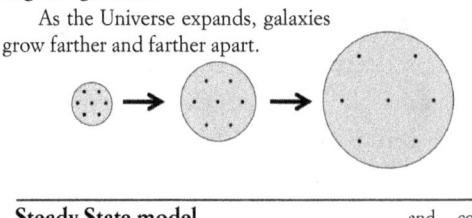

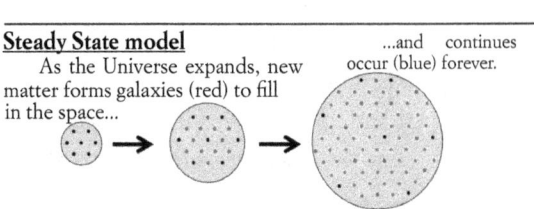

Figure 9.1. In the Big Bang model, matter gradually moves farther apart as it is dragged along by expanding spacetime, leaving increasingly large voids between structures. The Steady State model allowed for spacetime expansion, but devised a way for matter to continuously be generated to form new structures in the newly created voids.

The farther away we look, the longer the light has needed to travel to reach our eyes. Therefore, the farther away we look, the farther back in the past that light was emitted. Looking at great distances means we are seeing the Universe as it appeared farther and farther back in the past. According to the Steady State model, the distant Universe should look pretty much like the local Universe.

On the other hand, the Big Bang model predicted something drastically different. If the Universe began in a moment of creation, then it should look quite different in the past than it does now. If astronomers could see far enough away, the Big Bang model predicted that they would see a Universe that looked younger—or, at the very least, "old" galaxies would not exist and only young galaxies or young stars would be seen.

For the time being, the Steady State model gained little traction because the observations to support it simply weren't there. Surveys of the distant Universe were only in their infancy.

9.2 The big bang model develops

Meanwhile, the Big Bang model grew in popularity, driven by a Soviet physicist from Ukraine named George Gamow. Gamow studied nuclear physics in the Soviet Union before defecting to America in 1934. With his interest in this area, he began researching the mechanisms by which such a hot beginning could ultimately produce the atoms and elements we see today.

Astronomers had been able to measure the abundances of various elements in the solar system by using both meteorite samples and solar spectroscopy. The abundances in meteorites represented the material left over following the formation of the solar system, and they nicely matched up with the abundances found by studying the Sun's spectrum. This abundance pattern is shown in **Figure 9.2**.

As you can see in the figure, hydrogen is the most abundant element in the solar system—and in the Universe—with helium coming in second. While the remaining elements and their relative abundances now tell astronomers remarkable details about the way that elements are formed in stars, Gamow was interested in hydrogen and helium. Specifically, he wanted to know if the initial conditions after the Big Bang were such that hydrogen and helium could be created.

Lemaître had posited that the Universe began with all of the currently existing matter packed into a single, supermassive atom. He then figured that this atom began sequentially breaking apart into smaller and smaller atoms until the Universe arrived at its current configuration of tiny atoms (**Figure 9.3**). On the other hand, Gamow viewed the early moments of the Universe to be full of hydrogen atoms only. He theorized that conditions may have been ideal for some of that hydrogen to fuse together to form helium and the rest of the periodic table.

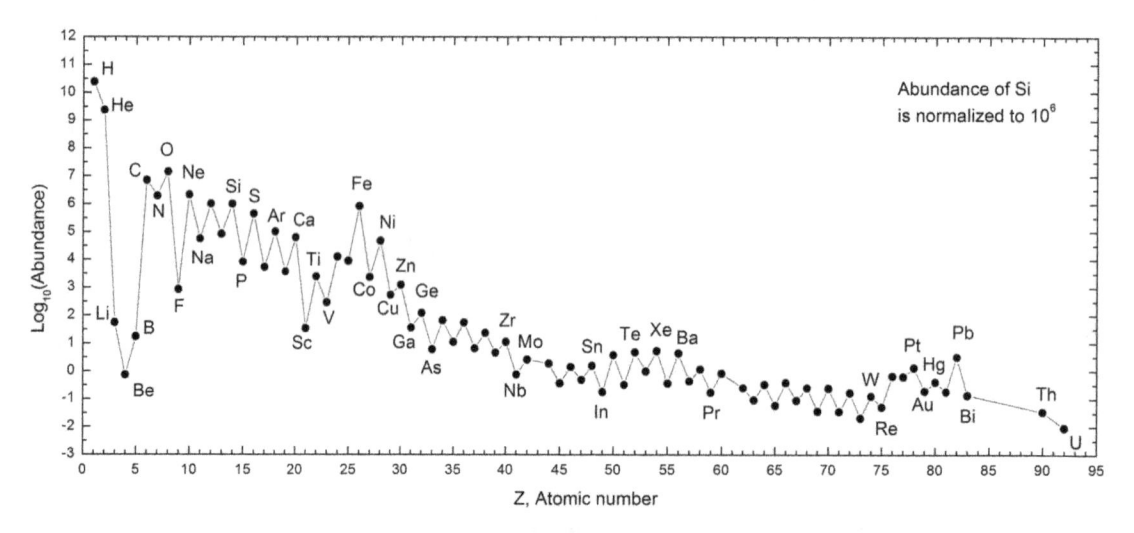

Figure 9.2. The solar abundance pattern. The horizontal axis counts off the number of protons in each element's nucleus, while the vertical axis corresponds to the log of the abundance value. For example, the abundance of lithium is roughly $10^2 = 100$. The log of 100 equals 2, hence its value on the axis. By expressing a factor of ten in scientific notation, taking the log of that number returns to you the exponent. Helium's abundance on this scale is over 1 billion. One billion = 10^9, and $\log(10^9) = 9$. What this boils down to is that while hydrogen doesn't look much more abundant than helium on this scale, there are actually about 10 times as many hydrogen atoms as helium atoms.

Copyright © 2007 by User:Orionus / Wikimedia Commons / CC BY-SA 3.0

a. Lemaître's model of the primeval atom

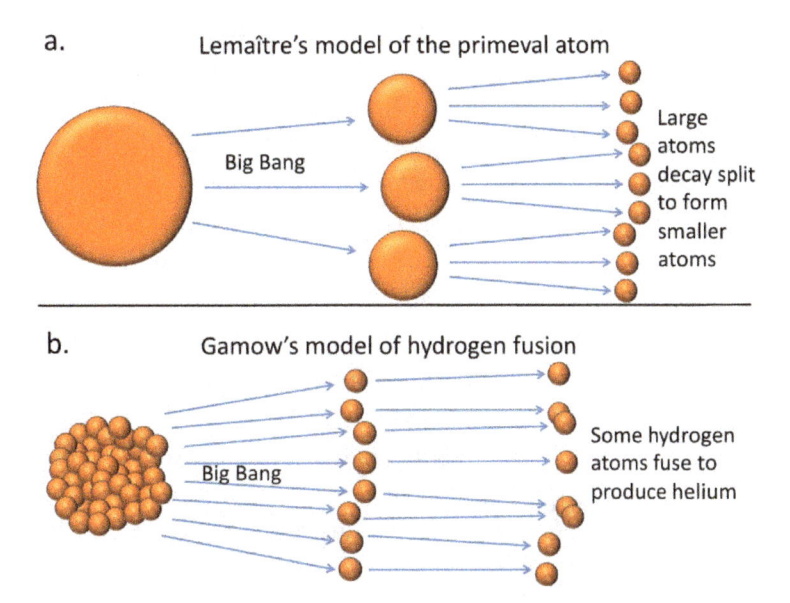

b. Gamow's model of hydrogen fusion

Figure 9.3. Contrast Lemaître's view of the creation of atoms with Gamow's model. (a) While Lemaître believed that all matter originated in a single, massive atom that subsequently broke apart into smaller atoms, (b) Gamow believed that everything started with small hydrogen atoms and built up from there. Gamow's model would ultimately be supported by observations.

By constructing a model that ran the cosmic expansion in reverse, Gamow was able to calculate how the density of the Universe would increase—and with it, the overall temperature—to see if the density and temperature were high enough to promote fusion. Together with his collaborators Ralph Alpher and Robert Herman, and using the starting assumption shown in **Figure 9.3**, they were successful in showing that there were roughly ten hydrogen nuclei for every one helium nucleus shortly after the Big Bang. This alone was a success for the Big Bang model, as it matched what had been measured in the Universe already (see **Figure 9.2**). According to their calculations, this should have taken roughly five minutes, at which point the temperature and density were too low for fusion to occur. While it wasn't exactly fair to call this result a prediction of the Big Bang model, since the relative abundances were calculated *after* the observation had been made, the agreement between theory and observation was encouraging.

From these calculations, Gamow, Alpher, and Herman came across one additional phenomenon that would prove to be a promising prediction. Running their model forward, once the helium nuclei had been created, the early Universe would have been full of free nuclei and free electrons, along with photons. The density was still high enough that these particles would have hindered the travel of light, thus causing the photons to frequently collide with

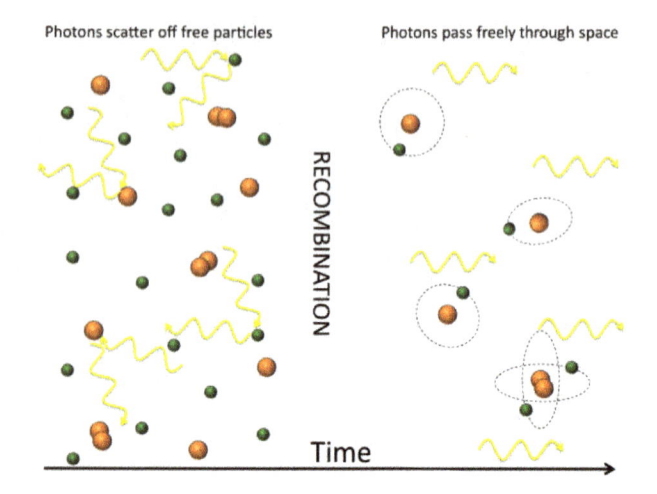

Figure 9.4. The epoch of recombination changed the Universe from opaque to transparent as electrons were bound up with nuclei to form atoms. Photons, no longer limited to repeated scattering, could now travel freely through space. This light is now believed to exist in the form of the cosmic microwave background.

these free particles and scatter. Because of this, light was unable to travel freely. Additionally, electrons could not combine with the nuclei to form atoms because the temperature was still too high—they were simply moving too fast for the electromagnetic force to enable the electrons to become bound to the nuclei. The Universe was said to be *opaque*.

According to their model, as time passed, and the Universe expanded and cooled, there came a point when the temperature dropped low enough that the electrons could finally bind to the nuclei, forming atoms. This event is called *recombination*, and is illustrated in **Figure 9.4**. At this point, the photons would interact with the atoms only if they had the right energy to be absorbed (recall Chapter 6). If their energy was not just right, they would pass right by the atoms, traveling unhindered through space. Suddenly, the Universe set loose a sea of photons that could now travel through space without being scattered. The Universe became *transparent* to light.

Gamow, Alpher, and Herman calculated that this light should now be present in the form of a background radiation. However, due to the cosmic expansion, this light should have been redshifted severely over time to the point where it would now be in the microwave/radio region of the electromagnetic spectrum. They published this prediction in the hope that an observer would prove them correct. Unfortunately, the technology related to detecting radio waves and microwaves was still quite new, and their prediction was all but forgotten. Nevertheless, the Big Bang model continued to increase in popularity.

9.3 Observational evidence—the cosmic microwave background

One could say that the development of radio astronomy as a field began in 1930 when a man named Karl Jansky, working for AT&T, built a moderately large radio antenna to attempt to map and measure a variety of sources of background radio noise. During several years of tests, he realized that some of these sources were not actually Earthly in nature. He first pinpointed the center of the Milky Way as one such source, and the field of radio astronomy was born. Thirty years later, thousands of celestial radio sources had been detected and studied. Even more interestingly, measurements and statistical calculations suggested that these objects, most of them what we now call radio galaxies and quasars, seemed to be more prevalent in the distant Universe—corresponding to earlier in the Universe's history. This result supported the Big Bang model, since the Steady State model would have predicted there to be just as many nearby radio galaxies as distant radio galaxies. The Steady State camp was losing its following as more astronomers crossed lines and joined the Big Bang side.

The nail in the coffin came when two men, Arno Penzias and Robert Wilson, made a serendipitous discovery. Working for AT&T's Bell Laboratories in 1963, the two men were using a large horn radio antenna to study celestial radio sources. In doing so, they found that regardless of where the antenna was pointed, there seemed to exist a uniform background noise. Driven to eliminate this noise, the men carefully ruled out any local sources that might be causing it. The task even involved cleaning the entire antenna, which you can see based on **Figure 9.5** was no small chore due to its size. A family of pigeons was discovered living inside the antenna and the job of cleaning up the white residue these birds left behind fell upon the men as well.

After a year, the antenna was cleaned and all Earthly and known celestial radio sources had been identified and accounted for—yet the background noise remained. Embarrassed at their inability to solve the problem, they gave up on it. However, as luck would have it, Penzias discussed the matter with a colleague who informed him of a paper postulating the existence of such background radiation as a relic of the Big Bang. While it was not their paper that was referenced, Gamow, Alpher, and Herman did not hesitate to point out that they too had predicted the existence of the cosmic microwave background (CMB). News of the discovery spread like

Figure 9.5. The horn antenna with which Arno Penzias and Robert Wilson (pictured) discovered the cosmic microwave background radiation.

NASA / Public Domain

Side-by-side comparison of testable predictions

Big Bang model	Steady State model
Nearby galaxies should appear older, more distant galaxies should appear younger.	An even mix of young and old galaxies should exist everywhere.
Because the universe began with high energy and density, a relic blackbody radiation from this should exist, which has since been redshifted to low energies.	With no beginning of the universe, the background radiation of the universe should only be from stars.
Light elements formed in the aftermath of the Big Bang. In particular, we should see no further creation of hydrogen.	Light elements form continually.

Result: The predictions of the Big Bang model were much more strongly supported by observation. As a result, the Steady State model faded from view and the Big Bang model became highly favored.

wildfire throughout the scientific community, and in 1978 Penzias and Wilson were awarded the Nobel Prize for Physics. The Steady State model, unable to provide a viable explanation for the existence of the CMB, gradually faded away. The Big Bang model was now mainstream. Although Fred Hoyle was an incredible scientist, he never accepted that the observational evidence did not support his idea and continued attempting to rework the Steady State model until his death in 2001.

But what makes the CMB so important to the Big Bang model? There are actually four reasons the CMB's discovery has been so widely heralded. First, its very existence as uniform relic radiation from the early evolution of the Universe supported the details of the model. Second, the radiation had the right wavelength. One can calculate the temperature of the Universe at the time the CMB was created because it is possible to calculate the temperature at which hydrogen atoms are ionized. If this represents the temperature at which electrons are ejected from hydrogen atoms, then slightly cooler than this represents the approximate temperature at which those electrons could recombine with nuclei to form atoms. Using Wien's Law (Chapter 6), the peak wavelength of the radiation present at that time can be calculated. Then, based on the expansion rate of the Universe, one can determine how much that light will be redshifted by now. The CMB was predicted to exist around a wavelength of 1 mm, and this agrees with observation.

The use of Wien's Law for that calculation relies on the third key point—the CMB's spectrum ought to be shaped like a blackbody (**Figure 9.6**). This is because the early Universe would have been quite dense, so any photons that existed would have been repeatedly scattered by free particles. Each collision changes the energy of the photons so that by the time they were released during recombination, the photons had a broad distribution of energies (wavelengths), with the peak wavelength corresponding to the average temperature of the matter via Wien's Law. If the CMB's spectrum was not shaped like a blackbody, this would have been disastrous for the Big Bang model. Additionally, the very fact that it *is* shaped like a blackbody means it can be attributed only to the Big Bang, since any other cosmic sources

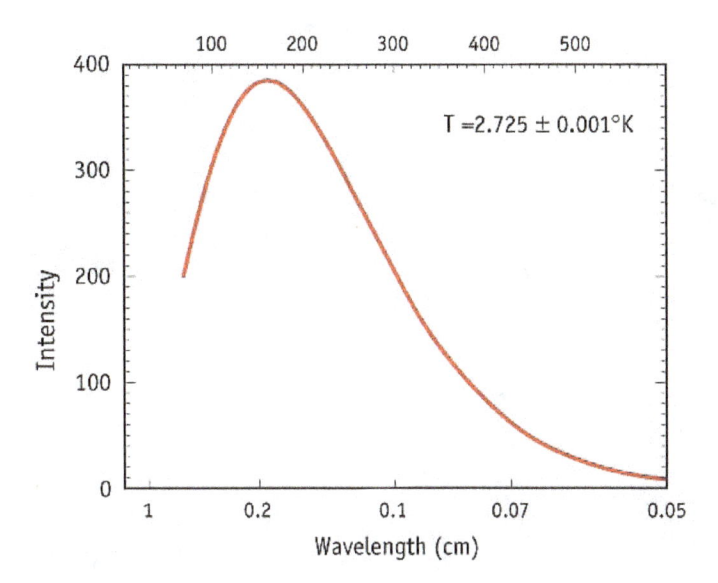

Figure 9.6. The CMB exhibits a nearly perfect blackbody spectrum, indicating that it originated from a hot, dense source.

NASA / Public Domain

would produce spectra that have absorption features, emission features, or a different shape entirely. It would be extremely hard for the Big Bang to explain a CMB that is not shaped like a blackbody, and it is very hard to scientifically explain a CMB that is shaped this way with a non–Big Bang model.

Over the decades since its initial discovery, the CMB has been carefully and repeatedly studied. While Penzias and Wilson first identified it as uniform over the entire sky (**Figure 9.7a**), it quickly became apparent that there must be some non-uniformity to it. The reason for this was that the radiation was created (or, more accurately, released) when free electrons combined with nuclei to form atoms. If the radiation was perfectly uniform, then that would imply the atoms in the early Universe were also perfectly uniform. A perfectly uniform distribution of atoms, though, would never coalesce to form structures like stars and galaxies because any force an atom feels pulling it in one direction would be canceled out by the same force pulling it in the opposite direction. Ultimately, this means that if the matter had been uniformly distributed early on, the Milky Way, the Sun, and the Earth would never have developed. We wouldn't be here.

More sensitive instruments were developed, and in 1989 the COsmic Background Explorer (COBE) satellite was launched into space. As you learned in Chapter 7, light from some regions of the electromagnetic spectrum is absorbed by the Earth's atmosphere and not easily detected from the ground. While some microwave-wavelength photons can reach the ground, many of them are absorbed by the atmosphere. To obtain the most accurate measurements of the CMB, it was necessary to put such detectors in space. The COBE satellite's results revealed

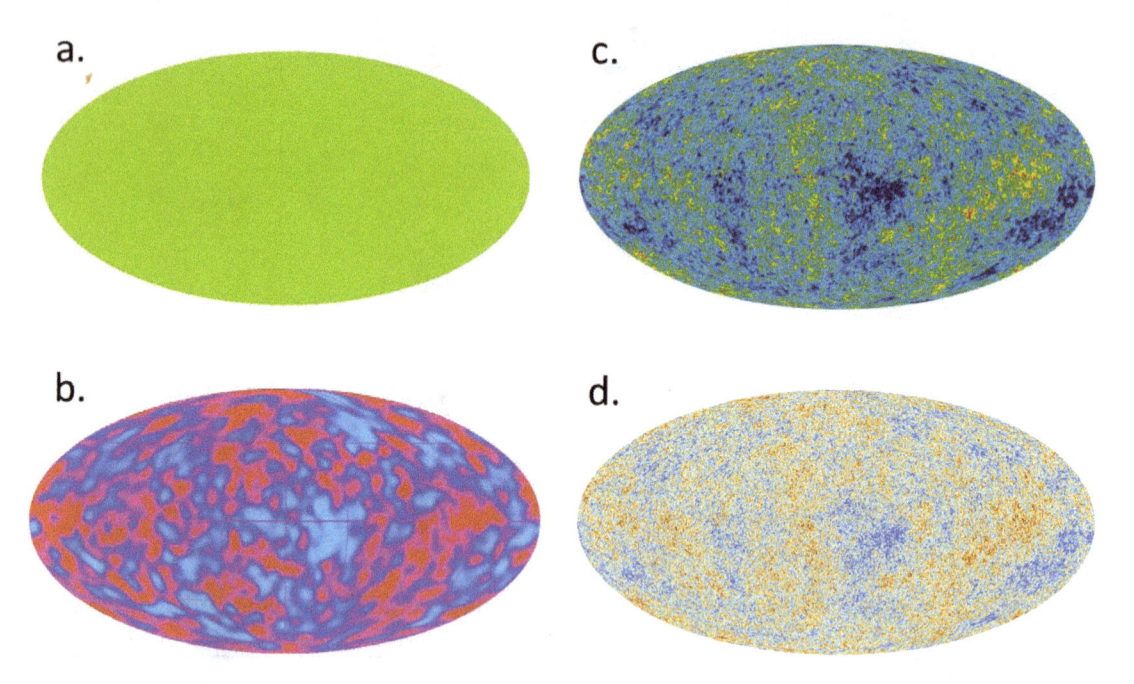

Figure 9.7. The cosmic microwave background as measured by four generations of instruments. (a) The Bell Laboratories radio antenna measured a uniform background, while (b) the COBE satellite was the first to map the full sky's fluctuations in CMB brightness—providing further support for the Big Bang model. (c) The WMAP satellite improved upon the resolution of these observations, providing incredibly precise measurements of various cosmological parameters. (d) The European Planck satellite mapped the CMB to greater precision and will almost certainly shed further light on the Universe through its results. It should be emphasized that the fluctuations seen in (c) and (d) are approximately at the level of 0.01%—extremely tiny!

(a) NASA / Public Domain (b) NASA Goddard Space Flight Center / Public Domain (c) NASA and WMAP Science Team / Public Domain (d) ESA and Planck / Public Domain

exactly what astronomers were hoping to see: there were extremely small fluctuations in the CMB—much smaller than were discernible by the Bell Laboratories radio antenna.

These fluctuations, at just a fraction of a percent, represented extremely minute temperature differences in the light from the CMB that corresponded to extremely minute density variations in the early Universe when the CMB was created. Areas where the temperature in the Universe was slightly cooler than average contained atoms that were moving just a bit slower than average. This allowed them to clump together just a bit and produce an area of a slightly higher density than average. The slightly cooler temperature released CMB photons that had energies slightly lower, corresponding to the blue regions in **Figure 9.7b**. Slightly hotter regions, where the density was a bit lower than average, produced photons with higher energies than average, corresponding to the red regions in **Figure 9.7b**. The primordial substructure that astronomers had hoped to see, and that the Big Bang model required, had been uncovered. For this discovery, the leaders of the COBE team were also awarded the Nobel Prize in Physics.

This non-uniformity in the CMB is the fourth reason that the detection of the microwave background was so important—it provides insight into the original seeds of structure.

Two decades later, the Wilkinson Microwave Anisotropy Probe (WMAP) resolved these fluctuations in remarkable detail (**Figure 9.8b**). Results from this satellite, to high precision, provided details about the geometry of the Universe, the age of the Universe, and values for the Hubble constant and other fundamental parameters of the Universe. Even more recently, the Planck satellite has produced an all-sky map of even higher resolution (**Figure 9.8c**), which will undoubtedly provide even greater insights into the nature of the Universe and conditions at the time the CMB was produced.

Observationally, the CMB represents the farthest back in time we can hope to see. The reason for this is the very same reason that clouds have the appearance of being solid. In an airplane, we fly through clouds all the time—they are certainly not solid. If you've ever been outside on a foggy evening, you've experienced what it is like to be inside a cloud. It does not seem all that different from being outside a cloud.

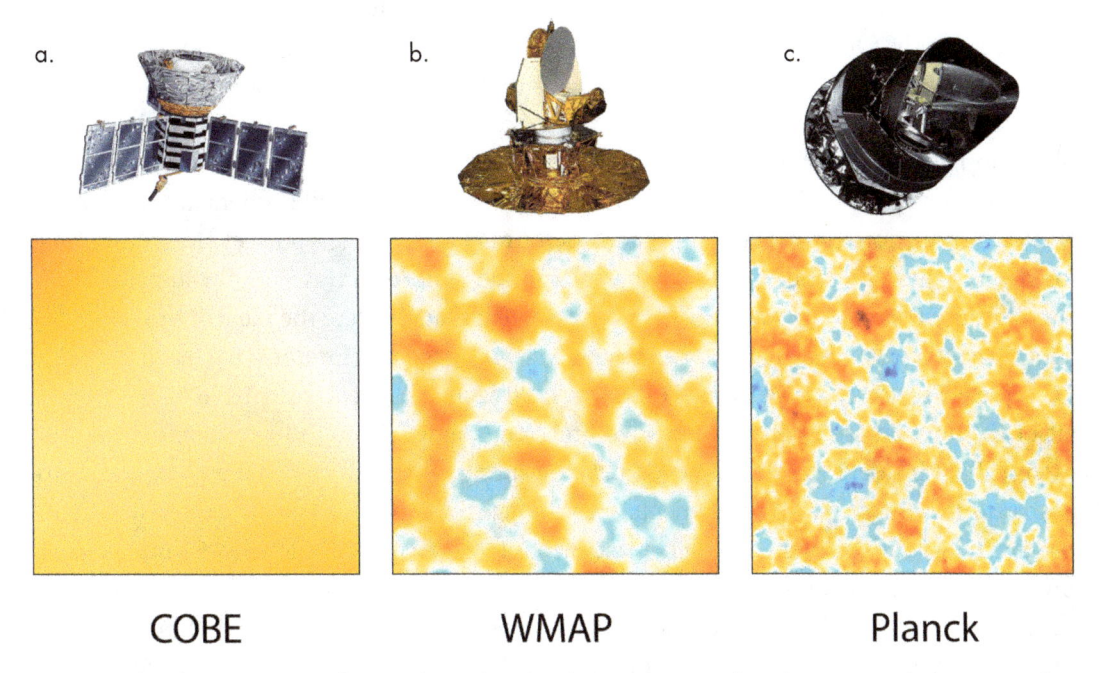

a. b. c.

COBE   WMAP   Planck

Figure 9.8. Each generation of space-based CMB observatory produced increasingly better resolution. (a) The COBE satellite was the first to resolve substructure in the CMB across the entire sky, while (b) WMAP and (c) Planck not only revealed a greater level of detail in this substructure but, as a result, also allowed astronomers to hone in on precise values of cosmological parameters. Knowing the values of such quantities as the Hubble Constant and the relative amounts of matter and energy allow astronomers to refine the Big Bang model even further.

NASA, JPL-Caltech, and ESA / Public Domain

Four reasons that the CMB is so important to the Big Bang model:

Uniformity	It appears everywhere in the sky at *very nearly* the same energy and brightness, indicating that its origin occurred everywhere in space at the same time and that the Universe was once compact enough for every point to be in thermal equilibrium.
Wavelength	Its spectrum has been redshifted due to cosmic expansion by an amount that corresponds to the amount of time since its origin.
Blackbody	Its blackbody shape indicates that it originated from a source that was both hot and dense, which correctly describes the early Universe.
Non-uniformity	The fact that the CMB isn't *perfectly* uniform provides evidence for fluctuations in the density of matter in the early Universe, which ultimately would have coalesced and grown to form all the structure we see.

According to photons of light, it's another story. Photons travel in straight lines through space (or through the air) until they meet an obstacle. That obstacle might be a wall, your eye, or even a particle. Generally, the particles in the air are not so dense that the photons notice them much. Collisions are rare. However, in the case of a cloud there exist small droplets of water suspended in vapor form that are just dense enough that a traveling photon encounters one rather quickly. Once it does, the photon is scattered and redirected. If you were standing on the other side of that water droplet, hoping to receive that photon, you'll be disappointed.

On a larger scale, photons represent an image of where that photon has been. Photons can come from light sources directly, or be scattered off something else. You can read this book because photons are being scattered off it. Some are absorbed, while others are not, and the way these absorptions and reflections occur determines the image you see. But you can't see through the page. Why? Because the paper blocks the light from traveling through it.

In a cloud, the density of water vapor is high enough that, just like the page of this book, light cannot travel through it before being scattered off a water droplet. Photons in a beam of sunlight are completely scattered away from their original direction of travel, so when a cloud moves in front of the Sun we can't see the Sun anymore. Some light, coming in from various angles, is scattered about inside the cloud and eventually escapes it. The final scatter it experiences before it is able to travel freely again represents the "surface" the cloud seems to have. We see that photon coming from the point at which it last encountered a water droplet. Taken together, those photons reveal to us what is sometimes called the "surface of last scatter," as shown in **Figure 9.9**.

In the case of the CMB, those photons were constantly being scattered about inside the dense soup of free electrons and nuclei. As electrons began binding together with nuclei to form atoms, the photons each made one last scatter off a free particle before finally being able to travel freely through space. We receive those traveling photons and see the point from which they originated—their "surface of last scatter." Beyond that point, the density was too high for light to make its exit—the Universe was opaque. Unfortunately, this means we can never directly observe the conditions that might have existed in the early Universe prior to the production of the CMB.

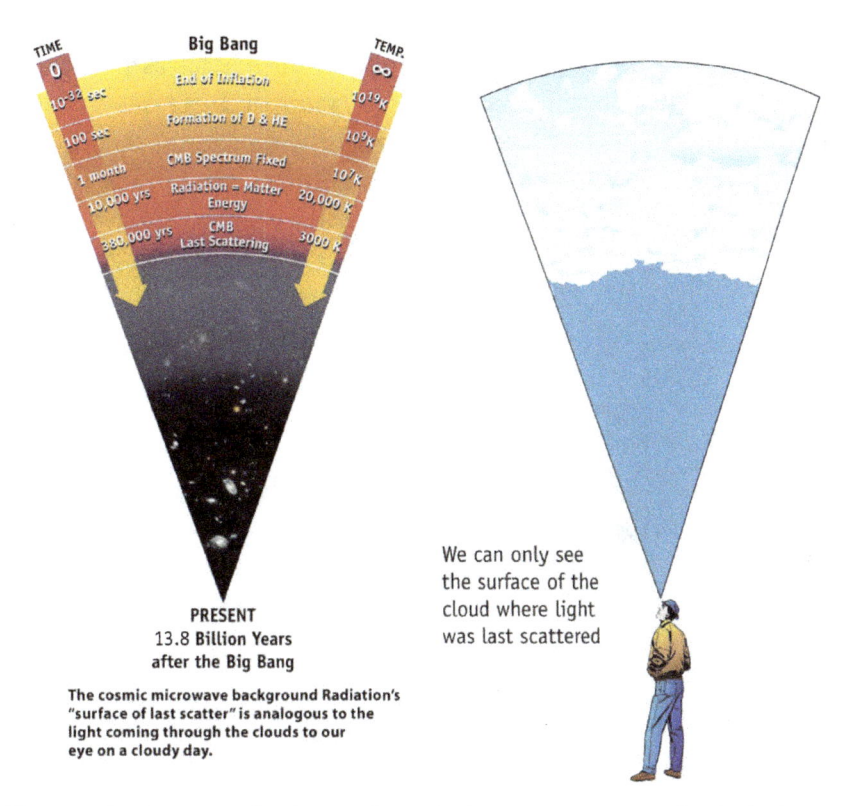

Figure 9.9. The cosmic microwave background acts in the same way as a cloud—both are opaque "surfaces" that do not allow light to travel directly through them from beyond. For this reason, we cannot hope to observe anything that may have occurred in the early Universe prior to the production of the CMB.

NASA and WMAP Science Team / Public Domain

9.4 The origin of matter

9.4.1 Matter and Antimatter

A couple questions naturally come to mind at this point: where did this matter originally come from, and what caused it to be slightly clumpy? To investigate this subject, we need to take a brief adventure into the fields of particle physics and quantum mechanics.

All of the substance in the Universe seems to be broken down into two overarching categories: matter and antimatter. We typically think of antimatter as being some sort of spooky stuff that belongs in science fiction movies that go straight to video. However, antimatter is

> **Quark**
> A fundamental particle of nature, of which there are six types, called *flavors*. Protons and neutrons are each made up of different combinations of three quarks.

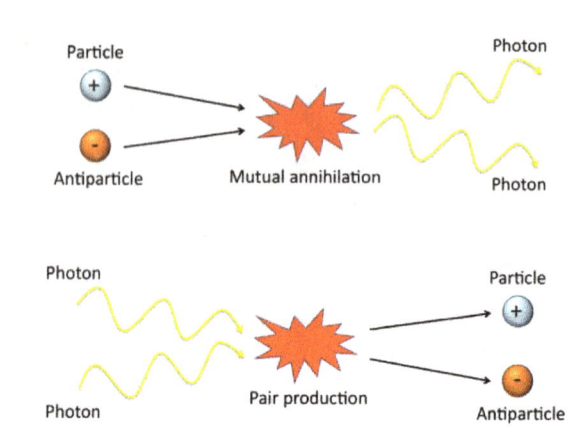

Figure 9.10. When matter and antimatter particles meet up, they annihilate each other to produce pairs of photons. Since photons are their own antiparticle, two identical photons can annihilate one another and produce a particle-antiparticle pair.

actually quite natural. It seems that every particle of matter has an antimatter partner—identical mass but opposite charge. An electron, for example, has a very low mass and a negative charge. The electron's antimatter partner, called a positron, is identical to an electron in mass but has a positive charge instead. Protons, and the *quarks* out of which they are made, also have antimatter partners. The weird thing about antimatter is that if a matter particle meets up with its antimatter partner, they annihilate each other and produce energy in the form of photons (**Figure 9.10**).

Some particles, like photons, are their own antiparticle. This means that two photons can actually annihilate one another as well, and when this happens a matter particle and its antimatter partner are produced.

The early Universe is believed to have been a very energetic place, with countless high-energy photons everywhere. Collisions between these photons would have been inevitable, and out of these collisions matter and antimatter particles would have been created.

Generally, these particle pairs should have been produced and destroyed in a symmetric way. However, for reasons that are still not well understood, it seems that more matter particles were created than antimatter particles. How do we know this? Because we are here! If the numbers had been equal, all matter would have annihilated all antimatter and nothing would be left but energy. Evidently, for every billion antimatter particles there were a billion and one matter particles—just enough so that all that was left over could create all that we see today. While particle physicists have observed particles spontaneously transform into antiparticles, and vice versa, this process should also be symmetric in both directions. The underlying reason for nature to seemingly prefer matter over antimatter remains a mystery.

Now, imagine you could use a microscope to zoom in on the very smallest size scales of space. These are size scales that make even atoms look as large as the observable Universe appears to us. At these truly tiny scales, the "smoothness" of space gives way to the uncertainty of quantum mechanics, as depicted in an artist's rendition in **Figure 9.11**. This means that at the smallest scales, fluctuations in energy can occur over the briefest of moments. These fluctuations can spontaneously produce pairs of matter and antimatter particles, which then immediately annihilate each other again.

An analog for this is a group of cyclists in the Tour de France. While the pack overall seems to be moving at a

Cosmologist

An astrophysicist who specializes in the study of the early Universe and its beginnings.

relatively uniform rate, within that pack individual cyclists might be passing others, or being passed. Similarly, an area of space might appear to have some average energy, but on the smallest scales, that energy is constantly fluctuating. Nowadays, these fluctuations are insignificant because they are tiny compared to the Universe. However, in the very distant past these fluctuations would have been large compared to the overall Universe because the Universe itself was extremely small too. It is believed that these types of fluctuations contributed to the initial non-uniformities in the Universe, which are now imprinted in the cosmic microwave background.

9.4.2 Dark Matter

You actually already know something about matter that gives insight into the fundamental matter distribution predicted by the Big Bang model. At an early age, you learned that in a gas, particles move about in such a way that the gas expands to fill its volume. You also know that if this gas is compressed, it tries to resist compression. This is why balloons get more difficult to inflate as they get larger. The air inside is becoming more compressed, since the rubber of the balloon doesn't continue to stretch as easily once it is already stretched out. This occurs at the basic level because the gas particles are colliding with one another, producing a gas pressure that resists further compression.

In the early Universe, the matter particles would have produced a similar sort of gas pressure, which should have smoothed out the distribution of matter quite quickly. So how, then, was structure still able to form? Particle physicists and *cosmologists* figure there must have existed a form of matter that did not behave in the same way—that is, it did not exert the same type of gas

Figure 9.11. Quantum fluctuations at the smallest size scales in space can spontaneously produce particle–antiparticle pairs that then immediately annihilate one another.

> **Dark matter**
> A different form of matter that does not interact electromagnetically, so it does not absorb or emit light. It can (so far) be detected only by its gravitational effects on normal matter.

Figure 9.12. The Coma cluster, a collection of over a thousand galaxies. It was the observed velocities of galaxies within clusters like this that first tipped off astronomers to the idea that there may be more there than meets the eye. Astronomers now believe there exists at least ten times as much dark matter within galaxy clusters as visible matter.

NASA, ESA, and the Hubble Heritage Team (STScI/AURA) / Public Domain

pressure against itself like "normal" matter does. This other form of matter should still exist today, but since it doesn't participate in processes related to thermodynamics or electromagnetism—particle collisions, heat exchange, light-matter interactions, etc.—it must not absorb or emit light. This matter has come to be known as *dark matter*.

As it turns out, it was not the Big Bang model that initially prompted speculation about the existence of dark matter, but observations of galaxies and galaxy clusters. In the 1930s, astronomer Fritz Zwicky was studying the velocities of galaxies within the Coma cluster—a cluster of galaxies shown in **Figure 9.12** that is located approximately 300 million light-years away. He noticed that the galaxies were moving much faster than they ought to be, given the amount of mass that was visible.

Examine **Figure 9.12** closely. Each of those objects is a galaxy within the cluster, and astronomers are able to infer the amount of visible matter based on the amount of light that is emitted. By adding up all the mass that is present, astronomers can then calculate how much gravitational force each galaxy should be experiencing within the cluster at its current location. The velocity that each galaxy has within the cluster corresponds to the strength at which it is being pulled by gravity—thus, if a galaxy is moving quickly, we know it is being pulled strongly. Zwicky's measurements indicated that the speeds of the galaxies were too high; the amount of mass that is visible just isn't enough to exert a gravitational force that was strong enough to produce such velocities. This implied to him that there must exist other matter, invisible matter, which is contributing its own gravitational influence.

As observations of individual stars in individual galaxies came in, including stars within the Milky Way, it became clear that stars were also moving "too quickly." When studying individual galaxies, astronomers can measure the rotational velocities of stars within spiral galaxies by using the Doppler effect. **Figure 9.13** illustrates how this is done. At a range of distances from the galactic center, astronomers pinpoint individual stars or even just a region of the spiral arm. The Doppler shift provides the velocity, and repeating this measurement provides measurements of the galaxy's rotational velocity as it depends on distance from the galactic center. The curve that is expected falls away at large distance, meaning that stars far from the center should be moving slowly because they are far from the central mass

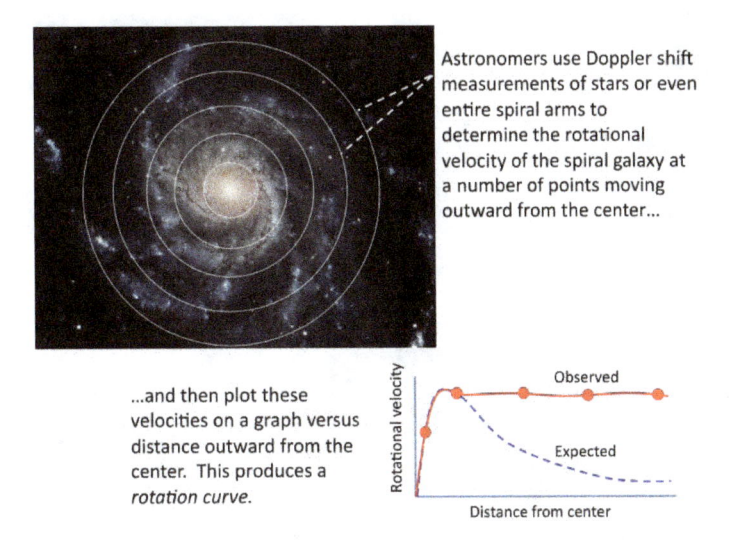

Astronomers use Doppler shift measurements of stars or even entire spiral arms to determine the rotational velocity of the spiral galaxy at a number of points moving outward from the center...

...and then plot these velocities on a graph versus distance outward from the center. This produces a *rotation curve*.

Figure 9.13. By measuring Doppler shifts within spiral galaxies, astronomers determined that stars within these galaxies are orbiting much faster than they should be if the visible matter alone was responsible for the gravity they experience. This led to the conclusion that dark matter must be present, providing the increased amount of gravitational force that creates such high orbital speeds.

NASA, ESA, K. Kuntz (JHU), F. Bresolin (University of Hawaii), J. Trauger (Jet Propulsion Lab), J. Mould (NOAO), Y.-H. Chu (University of Illinois, Urbana), and STScI / Public Domain

concentration. However, what is seen in nearly every spiral galaxy that has been measured is that the rotation curve actually flattens out. Stars at large distances from the center are orbiting at surprisingly high speeds. This must mean that there exists more matter—invisible dark matter—that is providing those stars with the added gravitational force needed to produce such high velocities.

The notion of some sort of mysterious, unseen substance existing in vast quantities rubbed many astronomers the wrong way. Perhaps, they suggested, our understanding of gravity is incorrect. Maybe Newton's formulation of gravity isn't quite right. Possibilities were explored for modifying Newtonian gravity to explain these galactic and stellar velocities. But more evidence continued to pour in.

While the stars are certainly an obvious component of galaxies and galaxy clusters, it turns out that there exist large amounts of gas in between stars and galaxies. This gas can be cool or it can be hot, depending on its environment. Within spiral galaxies, the gas tends to be cool, allowing it to collapse and form stars. Within elliptical galaxies, the small amount of gas present is much hotter. This is also true within galaxy clusters—the gas between the galaxies is heated to remarkable temperatures. This gas is heated due to the gravitational compression it is experiencing from the matter containing it. This means that the temperature of the gas is directly related to the mass present within the galaxy or galaxy cluster. **Figure 9.14** illustrates

Figure 9.14. The elliptical galaxy NGC 4649 as seen in optical (left; Hubble Space Telescope) and X-ray (right; Chandra X-ray Observatory) wavelengths. Both images are the same physical scale. The X-ray image reveals the presence of gas being heated due to the gravitational influence of dark matter permeating the galaxy.

(left) NASA and STScI / Public Domain (right) NASA, CXC, Univ. of California Irvine, and P. Humphrey et al. / Public Domain

Figure 9.15. The galaxy cluster Abell 1689. This image is a composite of Hubble Space Telescope and Chandra X-ray Observatory data, where the individual galaxies are easily seen embedded within a glowing halo of hot X-ray-emitting gas.

Source: https://commons.wikimedia.org/wiki/File:Abell_1689-_A_Galaxy_Cluster_Makes_Its_Mark_%28A_galaxy_cluster_at_a_distance_of_about_2.3_billion_light_years_from_Earth.%29_%282941478428%29.jpg

what this looks like in the elliptical galaxy NGC 4649, while **Figure 9.15** shows the same view in a galaxy cluster. In both cases, the rarified gas has been heated to such high temperatures that it emits X-rays.

When the amount of visible mass was added up, and its gravitational effects calculated, it became clear to astronomers that the amount of mass that was visible could not compress the gas to the temperatures seen in X-ray data. There simply had to be more matter present than was visible. Dark matter was the best explanation, but many people still remained unconvinced.

In came additional observations, this time relying on Einstein's general relativity. As you learned in Chapter 5, Einstein characterized gravity as being the result of an object accelerating along curved spacetime—curvature that was caused by the presence of large amounts of matter. Galaxy clusters, such as Abell 1689 shown in **Figure 9.16**, are some of the largest concentrations of matter in the Universe.

These large structures produce broad regions of curvature in their part of the Universe. As light rays travel through space from objects at larger distances from Earth, these light rays experience the curvature due to the galaxy cluster and their paths are deviated as a result. If the rays are deviated in the right way, they can eventually reach Earth—and our telescopes—even if the light rays weren't originally heading in that direction. Tracing these light rays back in a straight line, we have the impression that the image we see is actually extended above or below the galaxy cluster, and often in a way that is greatly distorted. **Figure 9.17** illustrates the principles that are occurring. This phenomenon, called gravitational lensing, occurs to various degrees depending on how much curvature exists in space, a factor that is related to the amount of mass present in the galaxy cluster.

Interestingly, this allows astronomers to measure the amount of mass that is present in a region of space simply by observing gravitational lensing features and modeling the mass distribution needed to recreate this. Newtonian gravity is not a factor at all, in this case. When astronomers run the modeling, they get a total mass enclosed

Figure 9.16. The galaxy cluster Abell 1689 contains thousands of galaxies and produces dramatic spacetime curvature. If you look closely, you can see curved and distorted images of distant galaxies being gravitationally lensed.

NASA, N. Benitez (JHU), T. Broadhurst (Racah Institute of Physics/The Hebrew University), H. Ford (JHU), M. Clampin (STScI), G. Hartig (STScI), G. Illingworth (UCO/Lick Observatory), the ACS Science Team and ESA / Public Domain

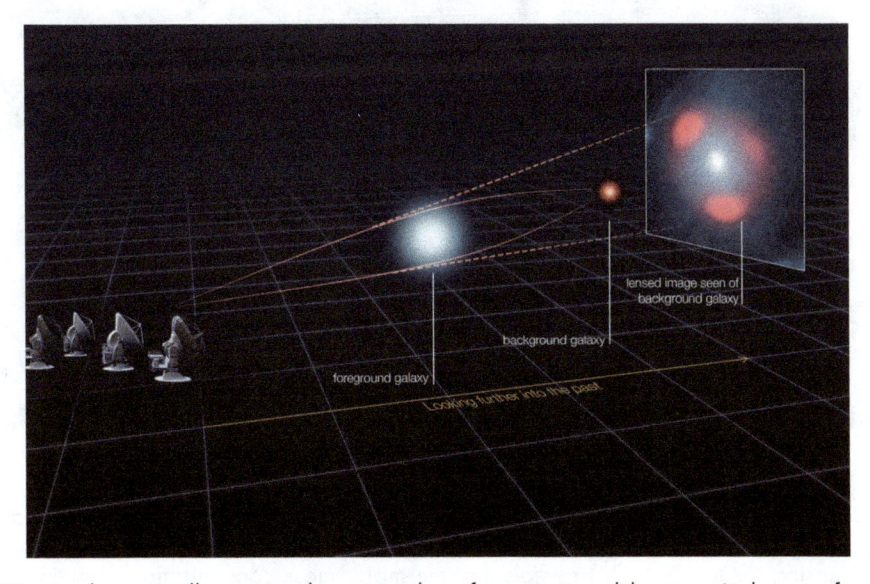

Figure 9.17. A schematic illustrating the principles of gravitational lensing. Light rays from a distant object are bent due to the curvature created by a massive structure between it and the observer. The observer sees the image of the distant object bent and distorted around the mass concentration, allowing the amount of mass present to be measured.

Copyright © 2013 by ALMA (ESO/NRAO/NAOJ), L. Calçada (ESO), Y. Hezaveh et al. / CC BY 3.0

within a given region of space. The calculated mass can then be compared to the mass that is visible by the light it emits. Again, there exists a huge discrepancy. The amount of mass needed to produce the observed gravitational lensing is much higher than the amount of visible mass present. There seems to be copious amounts of dark matter contained within galaxy clusters.

Most recently, theorists have used computer programs to model the distribution of matter within galaxy clusters based on extremely detailed measurements of gravitational lensing features. By combining these maps with optical images from the Hubble Space Telescope and X-ray data from the Chandra X-ray Observatory, it is now possible to investigate how the visible matter distribution compares to the overall matter distribution. The results, shown in **Figure 9.18**, are astounding.

The two galaxy clusters shown in **Figure 9.18** reveal two instances where galaxy clusters have collided with one another. Because galaxies themselves contain a lot of empty space, and are widely separated, the galaxy components of each cluster passed through one another and continued onward. The hot gas components of each collided and exerted gas pressure on one another, slowing down their passage through space and stripping them away from their galaxy components. Thus, the hot X-ray gas (shown in pink) is caught in between the two galaxy components now. However, gravitational lensing measurements indicate that the bulk of the

Figure 9.18. Computer-aided modeling of the mass distribution now allows astronomers to "see" dark matter. The images above show galaxy clusters MACS J0025.4-1222 (a; left) and 1E 0657-56 (b; right), both which are actually composed of two colliding galaxies, as they appear in optical (image of galaxies) and X-ray (pink). Shown in blue is actually where the majority of the mass—responsible for the gravitational lensing—is located. This reveals that the major mass concentrations are, in many cases, largely deficient in normal matter. Note: the pink and blue are only used to represent where the X-ray gas and dark matter are present—the colors themselves are not physical.

(left) X-ray (NASA/CXC/Stanford/S.Allen); Optical/Lensing (NASA/STScI/UC Santa Barbara/M.Bradac) / Public Domain (right) Sources: X-ray (NASA/CXC/CfA/M.Markevitch et al.); Optical (NASA/STScI; Magellan/U.Arizona/D.Clowe et al.); Lensing Map (NASA/STScI; ESO WFI; Magellan/U.Arizona/D.Clowe et al.)

matter is still surrounding the galaxies—not the hot X-ray gas. Astronomers can explain this only if the dark matter components of each galaxy cluster, which don't exert gas pressure on each other like the X-ray gas did, also passed through one another and remained with the galaxies. Thus, the blue regions represent the distribution of dark matter.

This is not unique to these two galaxy clusters; it is seen in others as well. It appears to be a very real result—and a result that ultimately put the seal of confidence on the existence of dark matter.

Based on these observations, made over the span of 80+ years, the case for the existence of dark matter is very strong. This is good news for the Big Bang model, because without dark matter's gravitational influence upon the hot gas believed to have been produced early on, structures could never have formed. Some astronomers might even say we owe our very existence to the presence of a particle about which we know almost nothing.

9.5 The model as we know it

With the groundwork laid, we can now look at the details of the Big Bang model as they were initially described by Gamow and Alpher and have been developed over the past 50–60 years. A large degree of scientific work and reasoning have contributed to making this model what it is today, but certain elements of uncertainty remain.

9.5.1 Planck Era ($t < 10^{-43}$ s)

Named for physicist Max Planck, this era represents a time during which we cannot describe the conditions of the Universe. The Universe would have been extremely dense at this point, which is usually where general relativity would be applied. However, length scales were also extremely small during this time, which is usually where quantum mechanics would be applied. Unfortunately, physicists do not currently have a theory that connects general relativity with quantum mechanics—the two models, used to describe two very different physical circumstances, do not play well together. It is possible that the entire Universe could have actually been packed into a volume that was infinitesimally tiny—called a singularity. However, the current Big Bang model does not require this to be true.

What physicists do believe they know is that the fundamental forces of nature were likely "united" into one superforce during this time. You are already familiar with two of the fundamental forces of nature: *gravity* and *electromagnetism*. There are two more that act on atomic-size scales, called the *strong nuclear force* and the *weak nuclear force*. The strong nuclear force is what binds quarks together into protons and neutrons, and also binds protons and neutrons together within the atomic nucleus. While it is the strongest force in magnitude, its range is small. The weak nuclear force causes unstable particles to change from one form to another

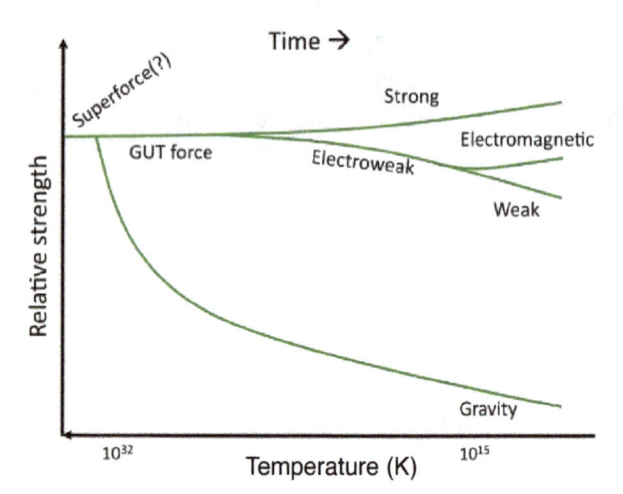

Figure 9.19. While we currently have four forces in our low-temperature Universe, they may have at one point in the early Universe been unified as one "superforce" when temperatures were extremely high.

and emit particles in the process. When an unstable atomic nucleus radioactively decays, this is the weak nuclear force at work. Its range is also small. On the other hand, while the electromagnetic force's strength is relatively high, it rarely builds up to noticeably significant levels because it is too easily canceled out by the presence of oppositely charged particles. Gravity, though it is the weakest of the four forces, is the one force that can effectively build up to tremendous total levels because it increases as more mass is added and while its strength drops with distance, it never goes to zero.

Figure 9.19 illustrates the sequence of events that are believed to have occurred to ultimately produce the four forces of nature. During the Planck era, the four forces may have been unified together, meaning that they were indistinguishable from one another. As the temperature cooled following the beginning of cosmic expansion, gravity is believed to have made itself distinct first—an event referred to as "freezing out." This marks the transition from the Planck era to the next era: the Grand Unified Theory (GUT) era.

9.5.2 GUT Era (t < 10^{-38} s)

Once the temperature was low enough for gravity to make itself known, the remaining force that is thought to have been left behind is known as the Grand Unified Theory force, or GUT force. The name is a bit of a misnomer, as technically the unified superforce would constitute more of a "unified theory," but sometimes terms just stick.

We cannot say much more about this era than we can about the previous era. The Universe was still quite compact and dense, and full of energetic photons. The temperature continued to drop as the Universe expanded until the point where the strong nuclear force froze out, leaving only the electroweak force behind. This event marks the next transition.

9.5.3 Electroweak Era (t < 10^{-10} s)

As the strong force froze out, the temperature was continuing to drop. Spontaneous particle pair production and mutual annihilation (**Figure 9.10**) was likely occurring at a frenzied rate. It may have been at this point that the matter–antimatter asymmetry began to appear, though this is still hypothetical. By the end of this era, temperatures had dropped enough that the electroweak force split into the electromagnetic and weak nuclear forces. The Universe now had four distinct fundamental forces of nature. Keep an eye on the clock—not even one nanosecond has elapsed yet!

One important thing to mention here is that the unification of electromagnetism and the weak force into an electroweak force has been experimentally verified. Particle accelerators in the 1980s were able to create environments with temperatures in the 10^{15} K range, which was hot enough to see particle signatures revealing the existence of the unified electroweak force. From here on out, the details of the Big Bang model become less speculative and more backed by theoretical calculations and observation.

9.5.4 Particle Era (t < 10^{-3} s)

As temperatures continued to drop with cosmic expansion, it eventually became too cool for spontaneous pair production to occur at significant levels anymore. Particles in existence at the time went through one final round of mutual annihilation and the matter particles produced by the matter–antimatter asymmetry were left behind for good. Quantum fluctuations had created regions where slightly more particles were located, and other regions with slightly fewer particles. These particles would have been free quarks and electrons, as well as other particles like neutrinos (particles that have nearly zero mass and travel at nearly the speed of light) and even more exotic particles. While we don't know how to describe them, dark matter particles were likely produced at this time as well.

Temperatures were initially still too high for these particles to bind together much, with a couple factors at play. The first factor is that a force can only act to bind two or more particles together if they come into close proximity with one another and are moving slowly enough that they don't "overpower" the strength of the force. In orbital dynamics (Chapter 4), if an object is moving faster than the escape velocity of a planet or star, that object will not remain bound by the gravitational force. Similarly, if quarks are moving too quickly, the strong force will not be able to bind them together.

The second factor at play is the energy of the surrounding sea of photons. In the event that particles happen to be bound together, if the average photon energy and density in the surrounding environment is high enough then any photons that collide with the bound particles will simply break them apart again. As the Universe cooled quarks gradually bound themselves together to form protons and neutrons.

9.5.5 Era of Nucleosynthesis (t < 5 min)

With continually dropping temperatures, particle velocities and photon energies dropped as well. Although the protons and neutrons were slowing down, they were still moving extremely fast—rapidly enough to collide and fuse together, but slowly enough for the strong nuclear force to hold them tight. The Universe transitioned into the Era of Nucleosynthesis

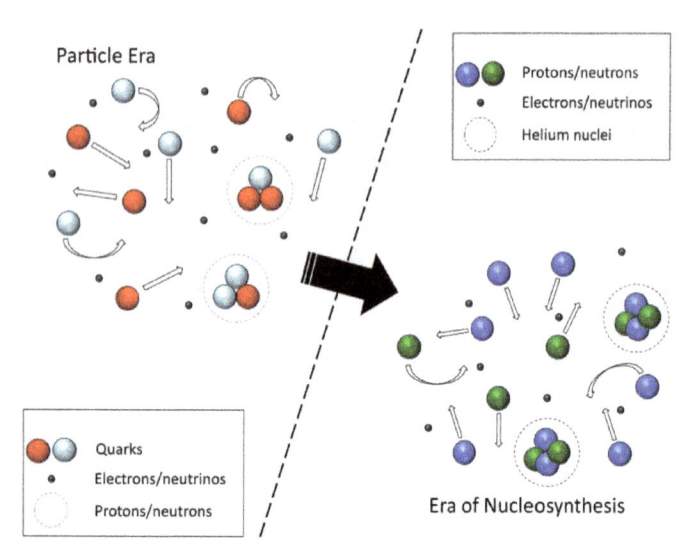

Figure 9.20. Cooling temperatures first allowed quarks to bind together into protons and neutrons, and then allowed the protons and neutrons to bind together to form helium nuclei.

(**Figure 9.20**), when helium nuclei began to be created. Remember that a hydrogen nucleus is simply a free proton, which had been present since the Particle Era.

Once the temperature dropped below about one billion K, these fusion reactions ceased and the *primordial abundances* were in place: approximately 75% hydrogen, 25% helium, and trace amounts of lithium. Out of this composition, the first stars would eventually form.

9.5.6 Era of Nuclei (t < 380,000 yrs)

Once nuclear fusion reactions had ceased, the Universe was filled with free electrons, free hydrogen and helium nuclei, neutrinos, dark matter, and photons galore. The speeds of these particles were still too high for the electrons to bind together with nuclei to form atoms, and photons were still sufficiently energetic to re-ionize any atoms that did form. The overall density was sufficiently high that photons continually collided and scattered with the free particles that existed everywhere. Since photons were unable to travel any great distance before being scattered again, the Universe was opaque to light.

After several hundred thousand years, the ambient temperature of the Universe had cooled to roughly 3,000 K. At this point, the free electrons began combining with the nuclei to form neutral atoms. The Universe suddenly opened up to the photons and light was finally able to travel unhindered through space. The light that would eventually become the cosmic microwave background was released.

9.5.7 Era of Atoms (t > 380,000 yrs)

We now exist in an Era of Atoms, where matter dominates over radiation and mighty galactic structures fill the Universe. After the end of the Era of Nuclei, the Universe when through what is referred to as the Dark Ages, when no stars yet existed and the only light around was the dimming cosmic background radiation. The gravitational influence of dark matter made clumpy by quantum fluctuations shepherded the normal matter gas into regions where the first stars could form, the first galaxies could develop, and galaxy clusters could grow. Successive generations of star formation produced increasing amounts of heavy elements that enriched interstellar space, eventually forming stars with relatively large amounts of heavy elements. These heavy elements would form planets. And, on at least one planet, life would emerge.

9.6 Problems and solutions

While the Big Bang model was reasonably successful at explaining the evolution of the Universe within a context of cosmic expansion, and was also able to explain the source of the CMB as well as predict the primordial helium abundance, it still left a few questions unanswered.

9.6.1 The Horizon Problem

Why is the cosmic microwave background radiation so uniform? While the recent space-based missions have done wonders for studying the minute fluctuations of the CMB, it is worth pointing out again that the CMB is actually remarkably smooth. Think of a well-worn basketball; those little bumps that were originally tacky to provide grip have been worn down to the point that it is close to being perfectly smooth. That represents the level of the fluctuations in the CMB. It's so smooth that you really need to look carefully to discern any non-uniformities.

Hold your arms out on each side of you, with your index fingers extended. Billions of light-years away in each direction, the CMB is very nearly the same temperature. But how can this be? In order for one location to "know" about another, there needs to be time for photons to be exchanged. This means that locations in space ought to be in equilibrium only if there has been time for photons, moving at the speed of light, to go from point A to point B. If two points are separated by 20 billion light-years, there simply hasn't been enough time for light to go from one to the other yet. In the early Universe, those disparate points would have needed time to come into thermal equilibrium with one another, and they wouldn't have had that kind of time in the standard Big Bang model before they were too widely separated to know about each other.

9.6.2 The Flatness Problem

In the early Universe, when space was extremely compact, the density of energy would have been tremendous. Albert Einstein told us that energy and mass are essentially two sides of the same coin—the same "stuff" but in a different form. This means that a concentration of energy then would have produced strong spacetime curvature in the same way a concentration of mass does today. According to the mathematics of cosmic expansion, any existing curvature should have become accentuated as dark matter pulled more normal matter to it.

General relativity indicates that there are three basic "geometries" that the Universe could have (**Figure 9.21**): it could be flat, where two parallel lines are always parallel; it could be positively curved like a ball, where two parallel lines eventually converge; or it could be negatively

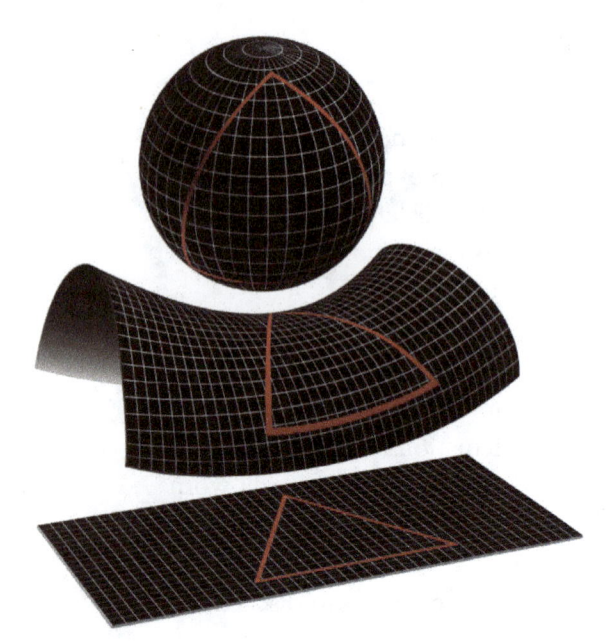

Figure 9.21. General relativity says that the Universe could have one of three different geometries that describe how light travels through the Universe. Our Universe appears to be flat.

curved like a saddle, where two parallel lines eventually diverge. Each geometry has a testable prediction by how it impacts the observed size of the fluctuations in the CMB, and the geometry of the Universe is, in general, determined by the density of matter and energy in the Universe. A large amount of matter and energy will produce positive curvature, a small amount of matter and energy will produce negative curvature, and "just the right amount" of matter and energy will produce a flat geometry.

One result from the WMAP satellite was that the Universe has a flat geometry*. In short, the sizes of the fluctuations observed in the CMB are, within the measurement uncertainty, exactly what was predicted by a Universe that was "just right." This means that the density of the Universe, expressed by the parameter Ω_0, appears to be equal to what is called the critical density (the density needed to make the Universe flat). This is actually quite convenient, because now when we add up all of the matter and energy that we can see and measure, we already know what the total should be! We will come back to this in Chapter 10.

So the question that comes to the mind of the budding scientist is: why should the Universe be flat? What occurred to cause the Universe to "choose" one of these geometries over the others? The Big Bang model did not have an answer to this question either.

9.6.3 Inflation

In the early 1980s, Alan Guth proposed and detailed a remarkable hypothesis. He had been studying the underlying physics of the transition between the GUT era and the Electroweak era. Based on the mathematics, he and his collaborators proposed that the very early Universe could have existed in a state called a *false vacuum*. You are probably already familiar with the concept of a vacuum, where space is empty or largely empty of matter. In a false vacuum, the

---

* Saying that the Universe is flat does NOT mean that it looks like the diagram in Figure 9.21. The Universe most certainly has three spatial dimensions within which we travel, so we can go up even though the diagram might not appear like it is allowed. All that we mean when we say that the Universe is flat is that if you headed off in any direction, the Universe would never wrap back on itself, bringing you back to where you started. You would just go on forever in a straight line.

Universe could have existed in a state that, much like an excited electron in an atom, was not at its lowest energy state for the ambient temperature at the time.

According to the inflation model (and its subsequent revisions by Andre Linde and Paul Steinhardt), a rapid temperature drop in the first instant after the Big Bang, which some have attributed to the breaking off of the strong nuclear force at the end of the GUT era, had left the Universe in a false vacuum state. This false vacuum state had a negative pressure, which acted to drive a moment of exponential inflation of space, whereby the Universe grew in volume by at least a factor of 10^{26} (**Figure 9.22**). As the Universe reached a lower vacuum energy state, this period of rapid expansion stopped after a split second and the more "normal" cosmic expansion resumed. The halting of inflation would then have caused the Universe to heat up again, and as it did so, it produced photons and other elementary particles.

The inflationary model has enjoyed a large amount of support because it effectively explains several things. First, it solves the horizon problem by allowing for a brief moment when the entire Universe was in *causal contact*—every location in the Universe was in sufficient proximity that it could come into thermal equilibrium, which ultimately led to a remarkably uniform CMB later on.

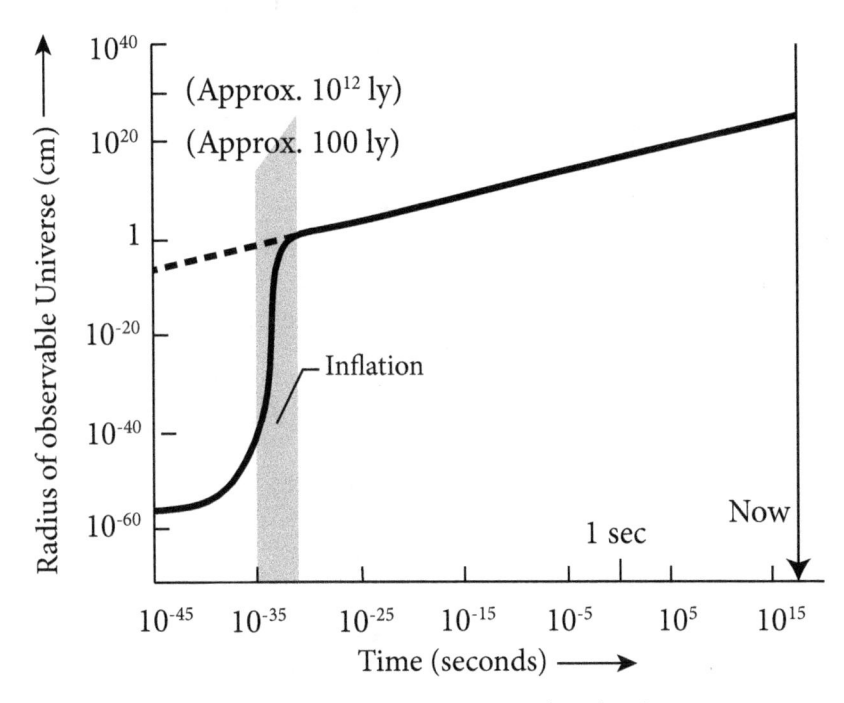

Figure 9.22. In the briefest of moments, the inflation model predicts that the Universe expanded by an incredible amount, producing a Universe that is now much larger than we can actually observe.

As **Figure 9.22** illustrates, had the Universe simply expanded at the same rate from the beginning until the present, reversing this process backwards along the dashed line leads to an initial size scale that was too large for the uniformity we observe—too large for the observable Universe in every direction to be in causal contact. If inflation did occur, then the Universe would have been much smaller originally (solid line), allowing the observable Universe to attain thermal equilibrium.

Second, it helps to explain the flatness problem because regardless of how much spacetime curvature existed prior to inflation, the amount of growth the Universe experienced would have ultimately stretched this curvature out so much that it would appear flat. Imagine you are an ant on the surface of a balloon as it is being inflated. From your perspective, the surface of the balloon appears increasingly flatter as time goes on. Similarly, our perspective on the Earth's surface is that the Earth's surface is flat. This is because we are tiny in comparison to the curvature. The curvature of the Universe appears to have been stretched out so much that it is either perfectly flat now or so close to being perfectly flat that we can't tell the difference.

Finally, inflation helped to explain the source of the density fluctuations that acted as the seeds of structure by taking the initial quantum fluctuations described in Section 9.4 and enhancing them to significant physical sizes. This allowed enough dark matter to accumulate to gradually pull the normal matter together to form stars and galaxies. Without this effect, it isn't clear that the gravitational force of tiny dark matter non-uniformities could have produced the structure we see today.

The Big Bang model on its own was a model that successfully helped astronomers explain the observed motions of galaxies and galaxy clusters in the Universe, the source of the cosmic microwave background, and the abundances of hydrogen and helium observed in the Universe. While it did not attempt to propose what caused the cosmic expansion to begin in the first place, by being supplemented with a viable description of exponential inflation in the very early moments following the Big Bang, the overarching model could explain nearly every observation that was tested against it. But, as is often the case, the Universe still had another trick up its sleeve.

CHAPTER TEN

While the Big Bang model itself has been worked out in a fair amount of detail and is widely accepted, models explaining the cause of the Big Bang are numerous. Many hypothesize the existence of extra spatial dimensions and parallel universes.

Copyright © 2012 by Depositphotos / Victor Habbick

Modern Cosmology

Learning Objectives

In this chapter you will learn:
- » The current census of the Universe
- » Why astronomers believe that cosmic expansion is accelerating
- » The most likely fate of the Universe
- » Hypotheses attempting to explain the cause of the Big Bang

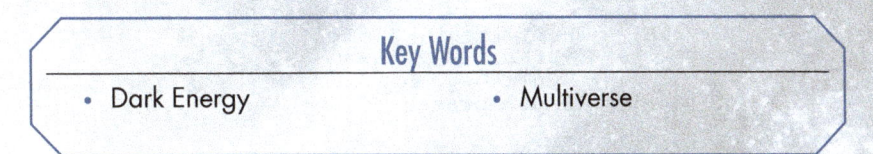

Key Words
- Dark Energy
- Multiverse

10.1 Accelerating expansion

The launch of the Hubble Space Telescope (HST) in 1990 promised the dawning of a new era in astronomy. Scientists and members of the public alike greatly anticipated the astounding imagery that such a space-based optical telescope could produce. Despite some initial hiccups, this floating observatory has, by all accounts, exceeded anyone's wildest expectations.

One of HST's primary mission objectives was to pin down the value of the Hubble Constant to within ten percent. To complete this mission, astronomers needed to collect data from the most distant galaxies that HST could see, which would require both high resolution and long CCD exposure times. Also on astronomers' minds was the goal of pinning down the deceleration parameter q_0—the rate at which the Universe's expansion was slowing down.

After collecting data from 31 spiral galaxies, the final results were published in 2000, shown in **Figure 10.1**. The team measured the Hubble Constant to be consistent with a value of 72 km/s/Mpc, meaning that for every megaparsec (million parsecs) farther in distance, galaxies are receding 72 km/s faster. To produce their results, the team leading this study utilized several different distance indicators. One of their most important distance indicators were exploding white dwarf stellar remnants, called Type Ia supernovae.

Many stars evolve in pairs called binary systems. As the two stars evolve, the more massive star evolves more quickly. It eventually swells up to become a *red giant* star, and if its mass is sufficiently low, then it gently blows off its outer atmosphere into space and leaves behind the hot core, a cinder known as a white dwarf. (**Figure 10.2**) There is an upper limit to the amount of mass this white dwarf can have: 1.4 solar masses.

As the second star evolves, it also eventually swells up as a red giant. These red giants can get so large that, if they are close enough to the companion star, they can actually transfer some of their mass onto the other star. In this scenario, the red giant dumps material onto the white dwarf. The white dwarf's mass gradually increases and it eventually exceeds the 1.4 solar mass limit. At this point, the interior pressure and temperature builds up to the point that the white

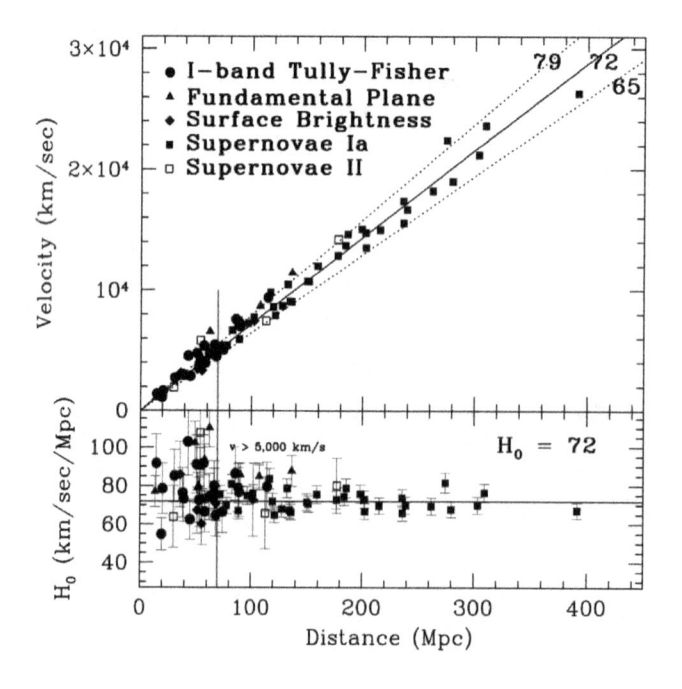

Figure 10.1. The Hubble diagram produced by years of data collection with the Hubble Space Telescope. Published in 2000, the team reported a value for the Hubble Constant of approximately 72 km/s/Mpc.

Source: Freedman et al., Astrophysical Journal, vol. 553, no. 1, p. 47.

dwarf undergoes runaway nuclear fusion reactions in its core. Without the original layers of gas on top of them, which were in place when the white dwarf was still a star, these fusion reactions go uncontained. The white dwarf explodes.

This type of supernova has been identified as a "standard candle," much in the same way that Cepheid variable stars have been. Since the white dwarf presumably explodes right when it exceeds the 1.4 solar mass limit, the amount of combustible "fuel" is the same in every case. If astronomers know how much fuel exists, then the energy output—the luminosity—can be calculated. All that is needed, then, is the apparent brightness of the supernova. By comparing this to its calculated luminosity, the distance to the supernova can be determined. These supernovae are extremely bright, allowing them to be seen at great distances, extending the Hubble diagram out significantly.

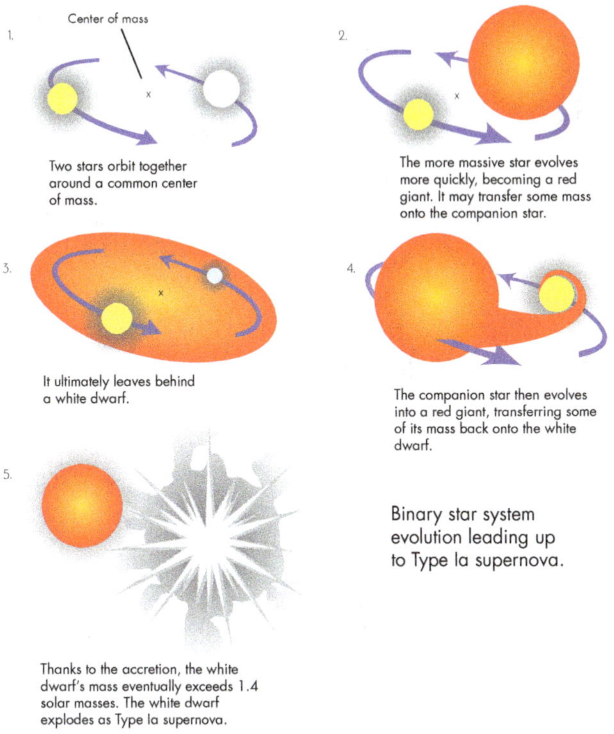

With the measured Hubble Constant of 72 km/s/Mpc, astronomers were then able to assign an estimate to the age of the Universe: 13.7 billion years. This age estimate was much more palatable than previous estimates because it allowed for a Universe that was older than the oldest stars in it. The most recent measurements have bumped this age estimate up to 13.8 billion years, while the Hubble Constant has fallen a bit to 70 km/s/Mpc. Things seemed to make sense.

Figure 10.2. Events leading up to the explosion of a Type Ia supernova.

Other astronomers had been studying this data set as well. In 1998, two independent teams of scientists published the results of their findings regarding the deceleration parameter. Astronomers had been expecting to measure how the rate of deceleration in cosmic expansion, thinking that the collective gravitational force from all the matter in the Universe would be slowing down the expansion. The results turned this expectation on its head.

Figure 10.3 shows the published data. The graph is actually a Hubble diagram, but its axes are flipped compared to what you have seen before (compare it to **Figure 10.1**). Do you see how the data points for the most distant galaxies turn upward? To see this more clearly, let's look at **Figure 10.4**, which shows the same curved trend plotted on a Hubble diagram as we are accustomed to looking at it, with distance along the horizontal axis.

> Astronomers now believe that the Universe is approximately 13.8 billion years old.

The vital point to remember as you look at **Figure 10.4** is that when we look at the most distant galaxies, we are seeing light that has been traveling for billions of years. This means that this light represents the Universe as it appeared a long time ago. In **Figure 10.4**, the right-hand side of the graph represents that time in the distant past, whereas the left-hand side represents the local Universe at the present time. The red lines drawn in indicate the slope of the line at both times. The slope of the line—the Hubble Constant, the expansion rate of the Universe—is flatter in the distant past than it is now. This means that the expansion rate of the Universe was lower in the distant past than it is now. Instead of measuring a deceleration parameter, astronomers discovered instead that the Universe's expansion is accelerating!

Astronomers were stunned. What sort of force would override the gravitational force of the entire Universe and speed up the rate of expansion? No known forces of nature could account for this behavior. The mysterious energy associated with such a repulsive force came to be known as *dark energy*.

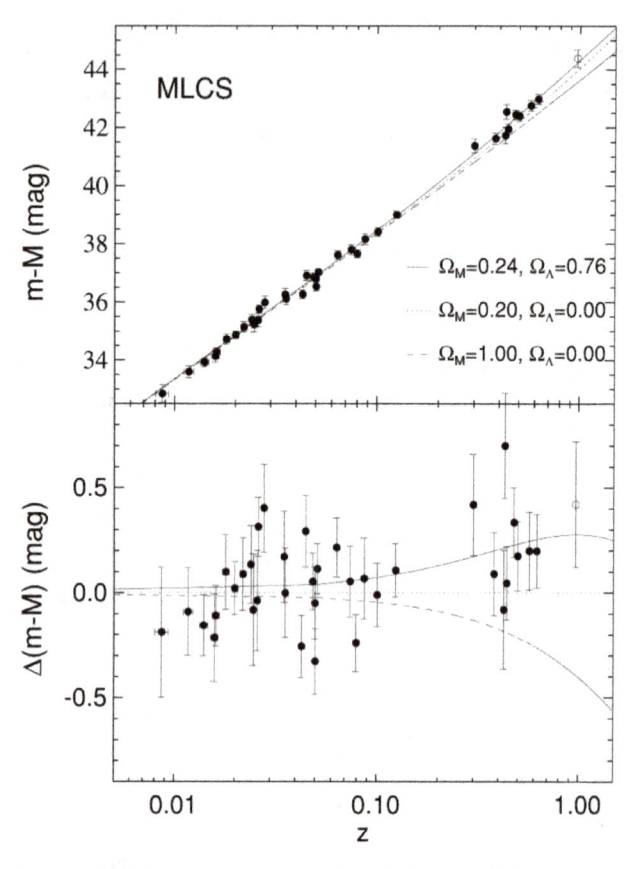

Figure 10.3. Data results from the Hubble Space Telescope, revealing that the slope of the Hubble diagram is not constant in time. At greater distances, corresponding to the more distant past, the slope is different.

Source: Riess et al., The Astronomical Journal, vol. 116, no. 3, p. 1022.

The Hubble "Constant," then, is really only a constant in space. This means that at any given moment, the expansion rate of the Universe is the same everywhere—currently about 70 km/s/Mpc. However, in the distant past this expansion rate appears to have been lower, and in the future it appears that it may be higher—it is not constant with time. What does this suggest about the future of the Universe? In order to consider the options, we must look at the overall composition of the Universe, because the amount of matter and energy that is present will determine just what the Universe ends up doing.

> **Dark energy**
>
> A repulsive force, the nature of which is unknown, that is causing the expansion of the Universe to accelerate—resisting gravity.

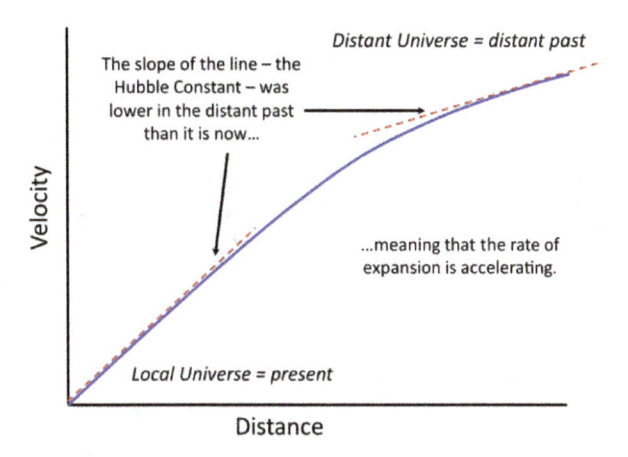

Figure 10.4. A representation of Figure 10.3 in a more recognizable way. The blue line represents the trend of the data points, and the red dashed lines indicate the slope of the data in the present (local) Universe and in the distant past.

10.2 Cosmic accounting

10.2.1 All we can see (and can't see)

Attempts at adding up all the matter in the Universe began much earlier than this, as astronomers battled with different solutions to Einstein's general relativity equations. Since his equations allowed for a number of different universes, it became necessary to try to pin down the parameters of our own Universe in order to figure out which solution was the most accurate.

As the Big Bang model has matured, it has become possible to make predictions about how much matter and radiation were present in the early Universe. This can be done because the amount of matter (both dark and normal) present would have impacted how quickly cosmic structures could form. If very little matter was present, then structure either wouldn't form at all or it would have formed very slowly. On the other hand, if a lot of matter was present, then structure would have formed quickly. Observing the distribution of galaxies and galaxy clusters, along with their sizes, allows astronomers to constrain the amount of matter that must exist.

With the discovery of dark energy, it is now important to consider how much it contributes to the overall mass-energy total of the Universe. Astronomers believe that dark energy must not have been a significant factor in the early Universe, or else it would have torn the Universe apart by now. What does seem clear is that at some point in the past something "turned on" dark energy's repulsive nature, causing the cosmic expansion rate to go from either constant or slowing down to speeding up.

Using observational results from the Hubble Space Telescope, as well as CMB data from the WMAP and Planck satellites, astronomers can input the values of various fundamental parameters into sophisticated computer models. These models output a "cosmic census" of the various mass and energy components, the results of which are shown in **Figure 10.5**. According to the best models to date, astronomers believe that the early Universe must have been dominated by dark matter. This needs to be true or, as mentioned earlier, structures like galaxies and galaxy clusters wouldn't have been able to form. Today, however, dark energy is by far the most dominant form of mass-energy in the Universe, with dark matter making up the bulk of the remainder. In fact, it appears that what we tend to consider "normal" matter is really less than 5% of all that is contained in the Universe. This may beg the question: what constitutes "normal"?

As cosmologists devised various solutions to Einstein's equations, they produced a few pictures of the potential future fate of the Universe (**Figure 10.6**). The discovery of dark energy added one more possibility, which now appears to be the most likely.

Closed Universe—In a Universe dominated by matter, the combined gravitational effects would quickly bring the initial cosmic expansion to a halt. This expansion would then reverse, causing the Universe to collapse back in on itself. In **Figure 10.6**, this corresponds to a line showing the Universe growing larger with time up to a certain level, and then turning back around and curving down to zero size again.

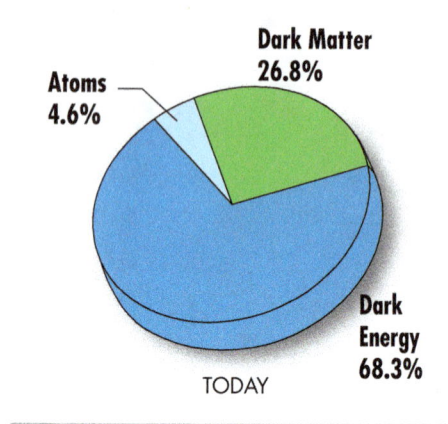

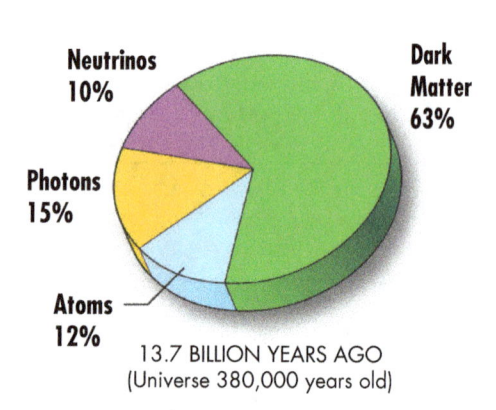

Figure 10.5. While the early Universe was dominated by dark matter, the Universe today is dominated by dark energy, with normal atomic matter composing less than 5% of the grand total.

NASA and WMAP Science Team / Public Domain

Critical Universe—If only matter is present, and that matter is just enough to produce a flat Universe on its own (where the matter density is equal to the critical density), then a critical Universe arises. In this scenario, the cosmic expansion continues forever but gradually tapers off over time. As time goes to infinity, the expansion rate continues slowing down toward zero.

Open Universe—If only matter is present but in quantities that are too low, then an open Universe results. In this model, the combined gravitational force of all the Universe's matter is too low to significantly impact the expansion rate of the Universe. The Universe continues expanding forever at a constant rate.

Accelerating Universe—Due to the presence of a repulsive dark energy force strong enough to overcome gravity, the expansion rate of the Universe continues to increase with time. Some models hypothesize a "Big Rip," where the acceleration continues to the degree where eventually stars, planets, and atoms are ripped apart and scattered. While it isn't known whether the Big Rip is in our future or not, the accelerating Universe scenario appears to be the most supported model according to current observations.

So what is this mysterious dark energy? At this point, astronomers don't know for sure. Many different models exist to describe it, but few additional observations have been made to begin to rule any of them out.

10.2.2 Cosmological Constant

One popular model describes it in terms of Einstein's cosmological constant. In this description, a volume of space has intrinsic energy—vacuum energy—and this energy produces a negative pressure throughout space. As space stretches out and creates more space, the magnitude of this pressure also increases, thus causing the expansion to accelerate. The appeal of this model is its simplicity: it is constant everywhere and simply compounds upon itself in a similar way that gravity compounds with increased mass. Being the most straightforward solution, it has been adopted by the current standard model of cosmology—called the Lambda Cold Dark Matter model (λCDM)—and enjoys a fair amount of predictive success. The biggest downside at the moment is that quantum mechanics predicts this vacuum energy should produce a cosmological constant about 120 orders of magnitude (factors of ten) higher than what has been measured.

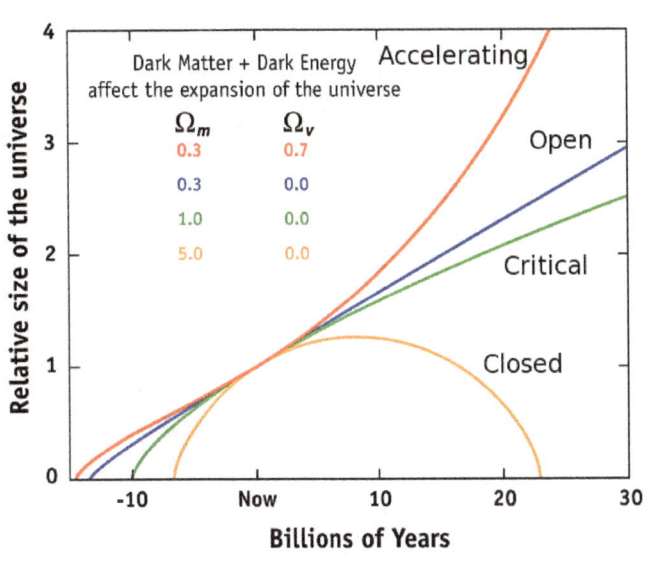

Figure 10.6. General relativity predicts several potential fates of the Universe. To everyone's surprise, the discovery of dark energy introduced a fourth—the accelerating Universe—which now appears most likely. In the figure, Ω_m represents the matter density of the Universe, while Ω_v represents the dark energy density. The point where all four lines meet corresponds to now because the models must agree on the current spacing between galaxies and the current value of the Hubble Constant.

NASA / Public Domain

10.2.3 Quintessence

A second model that has found a fair amount of support is a description where dark energy's strength varies in space and time. To describe such a force requires the presence of what is called a *scalar field*. Most fields with which we are familiar, such as gravitational fields and electromagnetic fields, are what are known as *vector fields*, where the strength of the field has both a magnitude and a direction. An object in such a vector field experiences a force due to the direction the vector field is pointing at that location. In a scalar field, there is only magnitude. For example, the number "3" is a scalar value; the description "3 blocks north" is a vector value—it specifies both magnitude and direction.

Scalar fields may seem exotic because we don't normally experience them in everyday life, but they are not uncommon in particle physics and quantum mechanics. Some may be real, while others are still hypothetical. The recent announcement of the discovery of the Higgs boson, a particle believed to be associated with any other particle's property of mass, may indicate the existence of such a scalar field called the Higgs field. The interaction of a particle with the scalar Higgs field would ultimately be what causes a particle to have a certain mass.

While quintessence is perhaps closer to being supported by quantum mechanical theory than the cosmological constant, it has not had any observational support at this time. Only time will tell which one of these models may be correct, and the solution may be a different cause entirely.

10.3 What may have caused the Big Bang?

If you were to ask anyone on the street for their biggest question regarding astronomy, the majority of those people would likely respond, "What caused the Big Bang?" It is only natural for us to ponder our existence, and since the Big Bang model is at this time the most widely accepted model for describing how the Universe has evolved, it's understandable to wish to know what set the whole thing in motion.

Unfortunately, as you learned in Chapter 9, not everything is within the grasp of science right now. Cosmologists believe they can trace back cosmic evolution to the Electroweak era with a high degree of certainty, and to the GUT era with a reasonable amount of confidence. However, going back even farther to the Planck era is currently beyond our capabilities. Until a viable theory of "quantum gravity" is developed, which combines elements of quantum mechanics and general relativity, cosmologists will work in partnership with theoretical physicists, particle physicists, and astronomers to develop models, refine hypotheses, and obtain observational data in the hopes of one day producing a complete model.

There is also this pesky business about time being so intimately related to space that when one runs cosmic expansion backward until all of space is compressed to a point, the flow of time also starts to look a bit funny. If the cause of the Big Bang is what brought space into being, then it is also what started the clock of time ticking. Asking questions like "What caused the Big Bang?" implies the existence of phenomena *before* the Big Bang, which are hard to describe when the word "before" loses its meaning. If time doesn't exist, can there be anything "before" that?

This all being said, these shortfalls have not prevented cosmologists from theorizing. Many promising models require the existence of additional spatial dimensions in the Universe. Within this setting, our 4-D Universe (three spatial dimensions, plus time) is just a smaller bit existing in a higher-order *multiverse* that consists of additional spatial dimensions. As an analogy, think of a thin sheet of paper, like tracing paper. The sheet is so imperceptibly thin that, for all intents and purposes, it is a two-dimensional object existing in a space of three dimensions.

> **Multiverse**
>
> A hypothetical space, consisting of more than three spatial dimensions, within which our Universe is embedded.

An ant walking along this sheet of paper is limited to going forward/backward and left/right. It can't go up because there is no up on the paper. Of course, this analogy breaks down because the ant itself is three-dimensional and therefore is quite familiar with the concept of up and down. A true two-dimensional creature would be just as flat as that sheet of paper and would be unable to even conceive of anything like up and down. Similarly, so the thinking goes, we are limited to our three spatial dimensions and are therefore unable to perceive any more than three spatial dimensions.

Several ideas have risen to the top as not-unreasonable guesses at the possible cause of the Big Bang. Keeping in mind that these are all nearly entirely speculative—and very difficult to study at all—let's consider a couple of these ideas.

10.3.1 Chaotic Inflation

This idea is also referred to as "eternal inflation" and consists of a multiverse of finite or infinite extent that makes the transition from false vacuum to true vacuum at different points at different times. Each of these separate regions would experience its own Big Bang, creating unique parallel "bubble universes" (see the figure at the opening of this chapter) that could, in principle, be physically disparate. Each would have its own set of physical laws and fundamental constants. Some might be potentially suitable for life; others might not. Some might expand forever, while others might end up closed and eventually collapse again.

10.3.2 Brane Multiverse

In an entirely different scenario, the multiverse is envisioned as being populated with two or more "membrane universes"—called *branes* for short, and illustrated in **Figure 10.7**. Within a higher-dimensional multiverse, each universe appears as a membrane of fewer dimensions. In this picture, these branes collide, perhaps even cyclically, and each collision marks the point of a Big Bang within each universe. The laws of physics within each universe are governed by the circumstances of the collision and may vary from one universe to the next. In an infinite multiverse, this scenario plays out endlessly.

These two examples are two of the more popular proposals in existence at the moment and really represent larger categories, within which are many variations with major or minor detail differences. While it may be impossible—or nearly so—to ever observationally study these

Figure 10.7. In the brane multiverse description, a potentially infinite number of low-dimension universes exist as membranes ("branes") within a high-dimension multiverse. Collisions between these branes produce events resembling the Big Bang within each brane.

Copyright © 2003 by Shutterstock / R.T. Wohlstadter

ideas in depth, astronomers can use details about the Universe in which we live to constrain the possibilities. For example, the search is on for a verifiable signature of so-called "B-mode polarization" in the cosmic microwave background that would potentially serve as a piece of direct evidence for inflation. Furthermore, each proposed scenario must be capable of producing a Universe like ours; if it cannot, it is excluded from consideration or modified.

10.4 Concluding remarks

While some of these notions may seem fantastical to you, many of the world's most brilliant minds have been working on the question of the Big Bang's cause for several decades. You may find it interesting to note that in the evolution of the Big Bang model, we have now arrived at two natural explanations for the cause of our Big Bang that both resemble a more grandiose version of the Steady State model—an eternally existing* multiverse producing universes for eternity. Thus, it seems that while George Gamow may have won the battle, perhaps Fred Hoyle won the war. Or, perhaps both are right in their own ways. Or, perhaps both are wrong and a new revelation is right around the corner.

This is the exciting element about astronomy, and science in general. Consistent, persistent study of a question often allows us to gain great insight and depth of knowledge about a subject or phenomenon. This same persistence can then, without warning, suddenly reveal something new that turns our understanding completely on its head. Such was the case with Galileo's observations of Jupiter and Venus, and such was the case with Einstein's publication of special and general relativity. Even more recently, this happened again with the discovery of cosmic acceleration.

This fact brings us back to the point made in Chapter 1: part of the reason science can never prove anything true is that the possibility always exists that the next observation will contradict the current paradigm and spur on a revolution in thinking. Scientists will always strive for that next observation or that next calculation that uncovers something new about the nature of the Universe—or the multiverse. In the meantime, we can rest assured that there will always be something new to learn, and live with the hope of seeing it uncovered during our lifetime.

---

* "It is worth noting that the scientific debate is still very much alive regarding whether or not an eternal sequence of "big bangs" within a multiverse is actually sustainable.

References

Chapter 1

1. Oort, J. H. "The structure of the cloud of comets surrounding the Solar System and a hypothesis concerning its origin." *Bulletin of the Astronomical Institutes of the Netherlands* 11. 1 (1950): 91-110.
2. Padmanabhan, T. *After the First Three Minutes: The Story of our Universe*. New York: Cambridge University Press, 1998.
3. Springel, V., et al. "Simulations of the formation, evolution and clustering of galaxies and quasars." *Nature* 435. 6 (2005): 629-636.

Chapter 2

1. Anthes, Rudolf. "Mythology of Ancient Egypt" in *Mythologies of the Ancient World*, edited by Samuel Noah Kramer, 443–472. Garden City, NY: Doubleday & Company, Inc., 1961.
2. Bodde, Derk. "Myths of Ancient China" in *Mythologies of the Ancient World*, edited by Samuel Noah Kramer, 443–472. Garden City, NY: Doubleday & Company, Inc., 1961.
3. Crowe, Michael J. *Theories of the World from Antiquity to the Copernican Revolution*. New York: Dover Publications, Inc., 1990.
4. Dicks, D. R. *Early Greek Astronomy to Aristotle*. Ithaca, NY: Cornell University Press, 1970.
5. Freund, Philip. *Myths of Creation*, New York: Washington Square Press, Inc., 1965.
6. Goetz, Delia and Morley, Sylvanus Griswold. *The Book of the People: Popol Vuh*. Los Angeles: Plantin Press, 1954.
7. Hoskin, Michael (Ed.). *The Cambridge Concise History of Astronomy*. New York: Cambridge University Press, 1999.
8. Krupp, E. C. *Echoes of the Ancient Skies*. New York: Harper & Row Publishers, 1983.
9. León-Patilla, Miguel. "Mythology of Ancient Mexico" in *Mythologies of the Ancient World*, edited by Samuel Noah Kramer, 443–472. Garden City, NY: Doubleday & Company, Inc., 1961.
10. Long, Charles H. *ALPHA: The Myths of Creation*, New York: George Braziller, Inc., 1963.
11. Pickover, Clifford A. *Archimedes to Hawking*. New York: Oxford University Press, Inc., 2008.
12. Singh, Simon. *Big Bang*. New York: Fourth Estate, 2004.
13. Ptolemaeus, Claudius. *Almagest*. Trans. G. J. Toomer. New York: Springer-Verlag, 1984.

Chapter 3

1. Armitage, Angus. *Copernicus: The Founder of Modern Astronomy*. New York: A. S. Barnes and Company, 1957.
2. Gade, John Allyne. *The Life and Times of Tycho Brahe*. Princeton, NJ: Princeton University Press, 1947.
3. Gingerich, Owen. *The Eye of Heaven: Ptolemy, Copernicus, Kepler*. New York: American Institute of Physics, 1993.
4. Gingerich, Owen. *The Book Nobody Read: Chasing the Revolutions of Nicolaus Copernicus*. New York: Penguin Books, 2005.
5. Gorst, Martin. *Measuring Eternity*. New York: Broadway Books, 2001.
6. Hawking, Stephen (Ed.). *On the Shoulders of Giants*. Philadelphia, PA: Running Press, 2002.
7. Hoskin, Michael (Ed.). *The Cambridge Concise History of Astronomy*. New York: Cambridge University Press, 1999.
8. Ferguson, Kitty. *Tycho & Kepler*. New York: Walker and Company, 2002.
9. Pickover, Clifford A. *Archimedes to Hawking*. New York: Oxford University Press, Inc., 2008.
10. Seeger, Raymond J. *Galileo Galilei, his life and his works*. New York: Pergamon Press, 1966
11. Singh, Simon. *Big Bang*. New York: Fourth Estate, 2004.
12. "The Derivation and First Draft of Copernicus's Planetary Theory: A Translation of the Commentariolus with Commentary," 1973, trans. N.M. Swerdlow, *Proceedings of the American Philosophical Society*, 117: 423–512.

Chapter 4

1. Cohen, I. Bernard, and George E. Smith, eds. *The Cambridge Companion to Newton*. New York: Cambridge University Press, 2002.
2. Gleick, James. *Isaac Newton*. New York: Pantheon Books, 2003.
3. Gorst, Martin. *Measuring Eternity*. New York: Broadway Books, 2001.
4. Hawking, Stephen (Ed.). *On the Shoulders of Giants*. Philadelphia, PA: Running Press, 2002.
5. Hoskin, Michael (Ed.). *The Cambridge Concise History of Astronomy*. New York: Cambridge University Press, 1999.
6. Maclaurin, Colin. *An Account of Sir Isaac Newton's Philosophical Discoveries.* New York: Johnson Reprint Corporation, 1968
7. Newton, Isaac, Ierome Bernard Cohen, and Anne Whitman. *The Principia. Preceded by a Guide to Newton's Principia by I. Bernard Cohen: Mathematical Principles of Natural Philosophy*. Berkeley: University of California, 1999.
8. Pickover, Clifford A. *Archimedes to Hawking*. New York: Oxford University Press, Inc., 2008.
9. Singh, Simon. *Big Bang*. New York: Fourth Estate, 2004.

Chapter 5

1. Bergmann, Peter. *Introduction to the Theory of Relativity.* New York: Dover Publications, Inc., 1976.
2. Einstein, Albert. "Does the Inertia of a Body Depend upon Its Energy Content?" 1905. *The Collected Papers of Albert Einstein. The Swiss Years: Writings, 1900-1909: English Translation.* Trans. Anna Beck. Ed. Peter Havas. Princeton, NJ: Princeton Univ. Pr., 1989. 172-74.
3. Einstein, Albert. "The Foundation of the General Theory of Relativity." 1916. *The Collected Papers of Albert Einstein. The Berlin Years: Writings, 1914-1917: English Translation.* Trans. Alfred Engel. Ed. Engelbert Schucking, Princeton, NJ: Princeton Univ. Pr., 1997. 283-339.
4. Einstein, Albert. "On the Electrodynamics of Moving Bodies." 1905. *The Collected Papers of Albert Einstein. The Swiss Years: Writings, 1900-1909: English Translation.* Trans. Anna Beck. Ed. Peter Havas. Princeton, NJ: Princeton Univ. Pr., 1989. 140-71.
5. Einstein, Albert. "On a Heuristic Point of View Concerning the Production and Transformation of Light." 1905. *The Collected Papers of Albert Einstein. The Swiss Years: Writings, 1900-1909: English Translation.* Trans. Anna Beck. Ed. Peter Havas. Princeton, NJ: Princeton Univ. Pr., 1989. 86-103.
6. Einstein, Albert. "On the Influence of Gravitation on the Propagation of Light." 1911. *The Collected Papers of Albert Einstein. The Swiss Years: Writings, 1909-1911: English Translation.* Trans. Anna Beck. Ed. Don Howard. Princeton, NJ: Princeton Univ. Pr., 1993. 379-87.
7. Einstein, Albert. "On the Movement of Small Particles Suspended in Stationary Liquids Required by the Molecular-Kinetic Theory of Heat." 1905. *The Collected Papers of Albert Einstein. The Swiss Years: Writings, 1900-1909: English Translation.* Trans. Anna Beck. Ed. Peter Havas. Princeton, NJ: Princeton Univ. Pr., 1989. 123-34.
8. Einstein, Albert. "On the Relativity Principle and the Conclusions Drawn from it." 1907. *The Collected Papers of Albert Einstein. The Swiss Years: Writings, 1900-1909: English Translation.* Trans. Anna Beck. Ed. Peter Havas. Princeton, NJ: Princeton Univ. Pr., 1989. 252-311.
9. Einstein, Albert. *Relativity.* New York: Three Rivers Press, 1961.
10. Gorbunov, Dmitry S. and Valery A. Rubakov. *Introduction to the Theory of the Early Universe: Hot Big Bang Theory.* New Jersey: World Scientific, 2011.
11. Gorst, Martin. *Measuring Eternity.* New York: Broadway Books, 2001.
12. Greene, Brian. *The Elegant Universe.* New York: Vintage Books, 1999.
13. Hafele, J. C., and Richard E. Keating. "Around-the-World Atomic Clocks: Predicted Relativistic Time Gains." *Science.* 177. 4044 (1972): 166-168.
14. Hafele, J. C., and Richard E. Keating. "Around-the-World Atomic Clocks: Observed Relativistic Time Gains." *Science.* 177. 4044 (1972): 168-170.
15. Hawking, Stephen (Ed.). *On the Shoulders of Giants.* Philadelphia, PA: Running Press, 2002.
16. Hawking, Stephen (Ed.). *A Stubbornly Persistent Illusion.* Philadelphia, PA: Running Press, 2007.
17. Isaacson, Walter. *Einstein.* New York: Simon and Schuster, 2007.
18. Ostriker, Jeremiah P. and Simon Mitton. *Unraveling the Mysteries of the Invisible Universe.* Princeton, NJ: Princeton University Press, 2013.

Chapter 6

1. Bohr, Neils. "On the constitution of the atom." *Philosophical Magazine* 26. 151 (1913): 1-25.
2. Bohr, Neils. "On the constitution of the atom. Part II – Systems containing only a single nucleus." *Philosophical Magazine* 26. 153 (1913): 476-502.
3. Bohr, Neils. "On the constitution of the atom. Part III – Systems containing several nuclei." *Philosophical Magazine* 26. 155 (1913): 857-875.
4. de Broglie, L. "Researches on the quantum theory." *Annales de Phisique* 3. (1925): 22-128.
5. Einstein, Albert. "On a Heuristic Point of View Concerning the Production and Transformation of Light." 1905. *The Collected Papers of Albert Einstein. The Swiss Years: Writings, 1900-1909: English Translation*. Trans. Anna Beck. Ed. Peter Havas. Princeton, NJ: Princeton Univ. Pr., 1989. 86-103.
6. Greene, Brian. *The Elegant Universe*. New York: Vintage Books, 1999.
7. Gribbon, John. *In Search of the Multiverse*. Hoboken, NJ: John Wiley & Sons, 2009.
8. Heisenberg, W. "On the Perceptual Content of Quantum Theoretical Kinematics and Mechanics." *Zeitschrift für Physik* **43. (1927)** 172-198. English translation in (Wheeler and Zurek, 1983), pp. 62-84.
9. Ostriker, Jeremiah P. and Simon Mitton. *Unraveling the Mysteries of the Invisible Universe*. Princeton, NJ: Princeton University Press, 2013.
10. Padmanabhan, T. *After the First Three Minutes: The Story of our Universe*. New York: Cambridge University Press, 1998.
11. Rutherford, E. "The scattering of α and β particles by matter and the structure of the atom." *Philosophical Magazine* 21. 125 (1911): 669-688.
12. Schrödinger, E. "Quantization as a problem of proper values (first communication)." *Annalen der Physik* 384. 4 (1926): 361-376.
13. Schrödinger, E. "Quantization as a problem of proper values (second communication)." *Annalen der Physik* 384. 6 (1926): 489-527.
14. Schrödinger, E. "Quantization as a problem of proper values (third communication: Perturbation theory with application to the Stark effect of the Balmer line)." *Annalen der Physik* 385. 13 (1926): 437–490.
15. Schrödinger, E. "Quantization as a problem of proper values (fourth communication)." *Annalen der Physik* 386. 18 (1926): 109–139.
16. Thomson, J. J. "Cathode Rays." *Proceedings of the Royal Institution* 9. (1897): 1-14.
17. Thomson, J. J. "On the structure of the atom: an investigation of the stability and periods of oscillation of a number of corpuscles arranged at equal intervals around the circumference of a circle; with application of the results to the theory of atomic structure." *Philosophical Magazine* 7. 39 (1904): 237-265.
18. Wheeler, J.A. and W. H. Zurek (Eds) *Quantum Theory and Measurement*. Princeton, NJ: Princeton University Press, 1983.

Chapter 7

1. Gorst, Martin. *Measuring Eternity*. New York: Broadway Books, 2001.

2. Ostriker, Jeremiah P. and Simon Mitton. *Unraveling the Mysteries of the Invisible Universe.* Princeton, NJ: Princeton University Press, 2013.
3. Pickover, Clifford A. *Archimedes to Hawking.* New York: Oxford University Press, Inc., 2008.

Chapter 8

1. de Sitter, Willem. "Einstein's theory of gravitation and its astronomical consequences." *Monthly Notices of the Royal Astronomical Society* 76. (1916): 699-728.
2. de Sitter, Willem. "Einstein's theory of gravitation and its astronomical consequences. Second Paper." *Monthly Notices of the Royal Astronomical Society* 77. (1916): 155-184.
3. de Sitter, Willem. "Einstein's theory of gravitation and its astronomical consequences. Third Paper." *Monthly Notices of the Royal Astronomical Society* 78. (1917): 3-28.
4. Einstein, Albert. "Cosmological considerations in the general theory of relativity." 1917. *The Collected Papers of Albert Einstein. The Berlin Years: Writings, 1914-1917: English Translation.* Trans. Alfred Engel. Ed. Engelbert Schucking, Princeton, NJ: Princeton Univ. Pr., 1997. 142-152.
5. Friedmann, A. "On the curvature of space." *Zeitschrift fur Physik* 10. (1922): 377-386.
6. Friedmann, A. "On the possibility of a world with constant negative curvature of space." *Zeitschrift fur Physik* 21. 1 (1924): 326-332.
7. Greene, Brian. *The Elegant Universe.* New York: Vintage Books, 1999.
8. Greene, Brian. *The Fabric of the Cosmos.* New York: Alfred A. Knopf, 2004.
9. Gorst, Martin. *Measuring Eternity.* New York: Broadway Books, 2001.
10. Hawking, Stephen W. *The Theory of Everything: The Origin and Fate of the Universe.* Beverly Hills, CA: New Millennium Press, 2002.
11. Hubble, Edwin. "NGC 6822, a remote stellar system." *Astrophysical Journal* 62. (1925): 409-433.
12. Hubble, Edwin. "A spiral nebula as a stellar system, Messier 33." *Astrophysical Journal* 63. (1926): 263-276.
13. Hubble, Edwin. "A spiral nebula as a stellar system, Messier 31." *Astrophysical Journal* 69. (1929): 103-158.
14. Hubble, Edwin. "A relation between distance and radial velocity among extra-galactic nebulae." *Proceedings of the National Academy of Science* 15. 3 (1929): 168-173.
15. Hubble, Edwin and Milton L. Humason. "The velocity-distance relation among extra-galactic nebulae." *Astrophysical Journal* 74. (1931): 43-80.
16. Leavitt, H. S. "1777 variables in the Magellanic Clouds." *Annals of the Harvard College Observatory* 60. (1908): 87-108.
17. Lemaître, Georges. "A homogeneous universe of constant mass and increasing radius accounting for the radial velocity of extra-galactic nebulae." *Annals of the Scientific Society of Brussels* 47. (1927): 49-59.
18. Lemaître, Georges. "Expansion of the universe, the expanding universe." *Monthly Notices of the Royal Astronomical Society* 91. (1931): 490-501.
19. Leavitt, H. S. and E. C. Pickering. "Periods of 25 variable stars in the Small Magellanic Cloud." *Harvard College Observatory Circular* 173. (1912): 1-3.

20. Ostriker, Jeremiah P. and Simon Mitton. *Unraveling the Mysteries of the Invisible Universe.* Princeton, NJ: Princeton University Press, 2013.
21. Padmanabhan, T. *After the First Three Minutes: The Story of our Universe.* New York: Cambridge University Press, 1998.
22. Pickover, Clifford A. *Archimedes to Hawking.* New York: Oxford University Press, Inc., 2008.
23. Silk, Joseph. *The Big Bang: The Creation and Evolution of the Universe.* San Francisco: W. H. Freeman, 1980.
24. Shapley, Harlow. "Studies based on the colors and magnitudes in stellar clusters. VII. The distances, distribution in space, and dimensions of 69 globular clusters." *Astrophysical Journal* 48. 10 (1918): 154-181.
25. Slipher, V. M. "The radial velocity of the Andromeda Nebula." *Lowell Observatory Bulletin* 2. (1913): 56-57.
26. Strömberg, G. "Analysis of radial velocities of globular clusters and non-galactic nebulae." *Astrophysical Journal* 61. (1925): 353-362.
27. Tyson, Neil deGrasse and Donald Goldsmith. *Origins.* New York: W. W. Norton, 2004.

Chapter 9

1. Alpher, R. A., Hans Bethe, and George A. Gamow. "The origin of chemical elements." *Physical Review* 73. 7 (1948): 803-804.
2. Alpher, R. A., Robert Herman, and George A. Gamow. "Thermonuclear reactions in the expanding Universe." *Physical Review* 74. 9 (1948): 1198-1199.
3. Alpher, R. A. and Robert Herman. "On the relative abundance of the elements." *Physical Review* 74. 12 (1948): 1737-1742.
4. Alpher, R. A. and Robert Herman. "Remarks on the evolution of the expanding universe." *Physical Review* 75. 7 (1949): 1089-1095.
5. Alpher, R. A.. "A neutron-capture theory of the formation and relative abundance of the elements." *Physical Review* 74. 11 (1948): 1577-1589.
6. Bondi, H. and Thomas Gold. "The steady-state theory of the expanding Universe." *Monthly Notices of the Royal Astronomical Society* 108. (1948): 252-270.
7. Bennett, C. L. "Nine-Year Wilkinson Microwave Anisotropy Probe (WMAP) Observations: Final Map and Results." *Astrophysical Journal Supplement* 208. 2 (2013): 20.
8. Dicke, R. H., P. J. E. Peebles, P. G. Roll, and D. T. Wilkinson. "Cosmic black-body radiation." *Astrophysical Journal* 142. 7 (1965): 414-419.
9. Gamow, George. "Expanding universe and the origin of elements." *Physical Review* 70. 7-8 (1946): 572-573.
10. Gamow, George. *My World Line.* New York: Viking Press, 1970.
11. Gorbunov, Dmitry S. and Valery A. Rubakov. *Introduction to the Theory of the Early Universe: Hot Big Bang Theory.* New Jersey: World Scientific, 2011.
12. Gorst, Martin. *Measuring Eternity.* New York: Broadway Books, 2001.
13. Greene, Brian. *The Elegant Universe.* New York: Vintage Books, 1999.
14. Greene, Brian. *The Fabric of the Cosmos.* New York: Alfred A. Knopf, 2004.
15. Gribbon, John. *In Search of the Multiverse.* Hoboken, NJ: John Wiley & Sons, 2009.

16. Guth, Alan. *The Inflationary Universe*. New York: Addison-Wesley Publishing Company, Inc., 1997.
17. Hawking, Stephen W. *The Theory of Everything: The Origin and Fate of the Universe*. Beverly Hills, CA: New Millennium Press, 2002.
18. Hinshaw, G., et al. "Nine-Year Wilkinson Microwave Anisotropy Probe (WMAP) Observations: Cosmological Parameter Results." *Astrophysical Journal Supplement* 208. 2 (2013): 19.
19. Hoyle, F. "A new model for the expanding Universe." *Monthly Notices of the Royal Astronomical Society* 108. 1 (1948): 372-382.
20. Hoyle, F., G. Burbidge, and J. V. Narlikar, J. V. "A quasi-steady state cosmological model with creation of matter". *The Astrophysical Journal* 410. 1 (1993): 437–457.
21. Impey, Chris. *How It Began*. New York: W. W. Norton & Co., 2012.
22. Ostriker, Jeremiah P. and Simon Mitton. *Unraveling the Mysteries of the Invisible Universe*. Princeton, NJ: Princeton University Press, 2013.
23. Padmanabhan, T. *After the First Three Minutes: The Story of our Universe*. New York: Cambridge University Press, 1998.
24. Penzias, A. A. and R. W. Wilson. "A measurement of excess antenna temperature at 4080 Mc/s." *Astrophysical Journal* 142. 7 (1965): 419-421.
25. Silk, Joseph. *The Big Bang: The Creation and Evolution of the Universe*. San Francisco: W. H. Freeman, 1980.
26. Singh, Simon. *Big Bang*. New York: Fourth Estate, 2004.
27. Smoot, G. F., et al. "Structure in the COBE differential microwave radiometer first-year maps." *Astrophysical Journal Letters* 396. (1992): L1-L5.
28. Tyson, Neil deGrasse and Donald Goldsmith. *Origins*. New York: W. W. Norton, 2004.
29. Weinberg, Steven. *The First Three Minutes*. New York: BasicBooks, 1988.

Chapter 10

1. Bennett, C. L. "Nine-Year Wilkinson Microwave Anisotropy Probe (WMAP) Observations: Final Map and Results." *Astrophysical Journal Supplement* 208. 2 (2013): 20.
2. Freedman, Wendy L., et al. "Final Results from the Hubble Space Telescope Key Project to Measure the Hubble Constant." *Astrophysical Journal* 553. (2001): 47-72.
3. Greene, Brian. *The Elegant Universe*. New York: Vintage Books, 1999.
4. Greene, Brian. *The Fabric of the Cosmos*. New York: Alfred A. Knopf, 2004.
5. Gribbon, John. *In Search of the Multiverse*. Hoboken, NJ: John Wiley & Sons, 2009.
6. Hawking, Stephen W. *The Theory of Everything: The Origin and Fate of the Universe*. Beverly Hills, CA: New Millennium Press, 2002.
7. Hinshaw, G., et al. "Nine-Year Wilkinson Microwave Anisotropy Probe (WMAP) Observations: Cosmological Parameter Results." *Astrophysical Journal Supplement* 208. 2 (2013): 19.
8. Impey, Chris. *How It Began*. New York: W. W. Norton & Co., 2012.
9. Mazzali, Paolo A., Friedrich K. Röpke, Stefano Benetti, and Wolfgang Hillebrandt. "A Common Explosion Mechanism for Type Ia Supernovae." *Science* 315. 5813 (2007): 825-828.

10. Ostriker, Jeremiah P. and Simon Mitton. *Unraveling the Mysteries of the Invisible Universe.* Princeton, NJ: Princeton University Press, 2013.
11. Padmanabhan, T. *After the First Three Minutes: The Story of our Universe.* New York: Cambridge University Press, 1998.
12. Perlmutter, S., et al. "Discovery of a supernova explosion at half the age of the Universe." *Nature* 391. 1 (1998): 51-54.
13. Perlmutter, Saul et al. "Measurements of Omega and Lambda from 42 high-redshift supernovae." *The Astrophysical Journal* 517. 2 (1999): 565-586.
14. Perlmutter, S., et al. "Measurements of the Cosmological Parameters Omega and Lambda from the First Seven Supernovae at z >= 0.35." *Astrophysical Journal* 483. (1997): 565-581.
15. Riess, Adam G., et al. "Observational evidence from supernovae for an accelerating Universe and a cosmological constant." *Astronomical Journal* 116. 3 (1998): 1009-1038.
16. Zlatev, I., Limin Wang, and Paul J. Steinhardt. "Quintessence, Cosmic Coincidence, and the Cosmological Constant." *Physical Review Letters* 82. 5 (1999): 896-899.

Index

A

absolute magnitude, 119–120
absorption spectrum, 121
adaptive optics, 114
Albert Einstein, 89
Anaxagoras, 29
antimatter, 155
aperture, 108
apparent magnitude, 118
Aristarchus, 30
Aristotle, 27
astronomical unit, 49

B

Big Bang, xi–xii, 9, 11–12, 122, 142–154, 156
blackbody spectrum, 102
blueshift, 132

C

Cepheid variable, 125, 128–129, 131, 175
chromatic aberration, 111
continuous spectrum, 101
cosmic microwave background, 9
cosmological constant, 139
cosmologists, 157

D

dark energy, 173, 176–179
dark matter, 158
Doppler effect, 132

E

eccentricity, 47–48
Edwin Hubble, 130
electromagnetic spectrum, 87
electromagnetism, 163
element, 64
ellipse, 47
emission spectrum, 100
Eratosthenes, 31

F

focal length, 109
Fred Hoyle, 144
frequency, 88

G

galaxy, 6
Galileo Galilei, 33, 51
George Gamow, 145
Georges Lemaître, 139
globular cluster, 12
gravity, 163

H
heliocentric, 30
Hubble's Law, 134
hydrostatic equilibrium, 128
hypothesis, 15

I
inflation, 169
interferometry, 113
Isaac Newton, 55
isotopes, 93

J
Johannes Kepler, 46

L
length contraction, 76
light-gathering power, 108
light-year, 3, 6
luminosity, 117

M
metals, 11
multiverse, 181

N
Nicolaus Copernicus, 41

O
objective lens, 110
observable universe, 8
orbital period, 48

P
parallax, 33
parsec, 120
photon, 68
Plato, 27
Ptolemy, 34

Q
quark, 156

R
recombination, 148
redshift, 132
reference frame, 71
Relativity, 17
resolving power, 108
retrograde motion, 34

S
science, 14
scientific notation, 5
semimajor axis, 48
spacetime, 80
spectroscopy, 99
spherical aberration, 112
Steady State, 144
Stefan-Boltzmann Law, 103
strong nuclear force, 93, 163

T
telescope, 110
theory, 16, 17
thermal radiation, 102
time dilation, 75
Tycho Brahe, 44

V
variable star, 128

W
wavelength, 88
weak nuclear force, 93, 163
Wien's Law, 104
William Herschel, 86